Cold Atmospheric Plasma-based Cancer Therapy (Second Edition)

Online at: https://doi.org/10.1088/978-0-7503-5537-7

IOP Series in Plasma Physics

Series Editor

Richard Dendy

Culham Centre for Fusion Energy and the University of Warwick, UK

About the series

The IOP Plasma Physics ebook series aims at comprehensive coverage of the physics and applications of natural and laboratory plasmas, across all temperature regimes. Books in the series range from graduate and upper-level undergraduate textbooks, research monographs and reviews.

The conceptual areas of plasma physics addressed in the series include:

- Equilibrium, stability, and control.
- Waves: fundamental properties, emission, and absorption.
- Nonlinear phenomena and turbulence.
- Transport theory and phenomenology.
- Laser–plasma interactions.
- Non-thermal and suprathermal particle populations.
- Beams and non-neutral plasmas.
- High energy density physics.
- Plasma–solid interactions, dusty, complex, and non-ideal plasmas.
- Diagnostic measurements and techniques for data analysis.

The fields of application include:

- Nuclear fusion through magnetic and inertial confinement.
- Solar–terrestrial and astrophysical plasma environments and phenomena.
- Advanced radiation sources.
- Materials processing and functionalisation.
- Propulsion, combustion, and bulk materials management.
- Interaction of plasma with living matter and liquids.
- Biological, medical, and environmental systems.
- Low-temperature plasmas, glow discharges, and vacuum arcs.
- Plasma chemistry and reaction mechanisms.
- Plasma production by novel means.

A full list of titles published in this series can be found here: https://iopscience.iop.org/bookListInfo/iop-plasma-physics-series.

Cold Atmospheric Plasma-based Cancer Therapy (Second Edition)

Michael Keidar
Department of Mechanical and Aerospace Engineering, George Washington University, Washington, DC, USA

Dayun Yan
Department of Mechanical and Aerospace Engineering, George Washington University, Washington, DC, USA

Li Lin
Department of Mechanical and Aerospace Engineering, George Washington University, Washington, DC, USA

IOP Publishing, Bristol, UK

ISBN 978-0-7503-5537-7 (ebook)
ISBN 978-0-7503-5535-3 (print)
ISBN 978-0-7503-5538-4 (myPrint)
ISBN 978-0-7503-5536-0 (mobi)

DOI 10.1088/978-0-7503-5537-7

Version: 20230901

IOP ebooks

British Library Cataloguing-in-Publication Data: A catalogue record for this book is available from the British Library.

Published by IOP Publishing, wholly owned by The Institute of Physics, London

IOP Publishing, No.2 The Distillery, Glassfields, Avon Street, Bristol, BS2 0GR, UK

US Office: IOP Publishing, Inc., 190 North Independence Mall West, Suite 601, Philadelphia, PA 19106, USA

Michael Keidar dedicates this book to his wife Victoria for her tremendous support and sacrifice allowing him to focus on research over many years.

He also dedicates this book to the blessed memory of his mother, Dina.

Contents

Acknowledgements

The authors acknowledge the contribution of all current and past members of the Micropropulsion and Nanotechnology Laboratory at the George Washington University.

Author biographies

Michael Keidar

Michael Keidar is a James Clark Professor of Engineering. His research concerns plasma physics and engineering with application in plasma medicine, advanced spacecraft propulsion, and plasma-based nanotechnology. He has authored over 350 journal articles, 300 conference papers, and is author of the textbook *Plasma Engineering* and the book *Plasma Cancer Therapy*. He received the 2017 Davidson award in plasma physics. In 2016 he received the AIAA Engineer of the Year award for his work on micropropulsion, resulting in the successful launch of nanosatellite with thrusters developed by his laboratory. He received the Plasma Medicine Award in 2021. Professor Keidar serves as an Editor-in-Chief of the *Journal of Electric Propulsion*, is the *AIP Advances* academic editor, Associate Editor of *IEEE Transactions in Radiation and Plasma Medical Sciences* and is a member of editorial board of a half dozen other journals. He is an elected Fellow of National Academy of Investors, Fellow of American Physical Society and Fellow of American Institute of Astronautics and Aeronautics. He is elected President of the Electric Rocket Propulsion Society.

Dr Dayun Yan

Dr Dayun Yan grew up in Xiaogan, Hubei Province, China. His father was a dean at a university. He obtained his MS degree (mechanical engineering) from Tsinghua University in China with an 'Excellent Dissertation' award. After that, he worked as a research assistant in the State Key Laboratory of Tribology (SKLT) for two years. He later obtained his PhD degree (mechanical engineering) at George Washington University in 2016 under the supervision of Dr Keidar. After that, Dayun started his postdoctoral training at the same lab. His current research focuses on several core problems in non-thermal plasma, plasma sources, and plasma medicine. Dayun has published 60+ papers in peer-reviewed journals with ~3000 citations. He wrote the first edition of *Cold Plasma Cancer Therapy*. Besides, he served as the discussion leader or session chair at important plasma-related conferences (GRS, ICOPS, and PPPS). Several presentations have been made by him at important plasma conferences (GRS, GRC, ICOPS, IWPCT, PPPS, ICPM, and ISPC). He was guest editor of four journals (*Applied Sciences*, *Pharmaceutics*, *Biomedicines*, and *Cancers*) and is a peer-reviewer for ~90 journals.

Li Lin

Li Lin completed his PhD program at the end of 2019 at George Washington University (GWU) and was a postdoctoral researcher at Micropropulsion and Nanotechnology Laboratory (MpNL) from January 2020 to January 2023. He is now a research scientist of MpNL at GWU. His research interests are low-temperature plasma physics and chemistry, especially the chemical pathway networks and adaptive control of the plasmas that are applied to biomedical and environmental science such as selective cancer therapies, microorganism deactivations, and air purifications. Li is also working on interdisciplinary topics such as combining machine learning technologies and neuromorphic concepts with numerical simulations and plasma diagnostics, such as optical emission spectroscopy for UV–VIS, Fourier-transform spectroscopy for infrared, Rayleigh microwave scattering, laser Michelson interferometer, and a high-speed camera. Li also provides hardware support for his colleagues in MpNL, such as developing new plasma generators. Li is currently serving for several academic journals including as an editorial board member of *Scientific Reports*, a topic editor of *MDPI Catalysts*, and a review editor of *Frontiers in Physics*, etc. Li also received the 'trusted reviewer' and 'outstanding reviewer' status from the Institute of Physics (IOP). Li is a member and has chaired conference sessions of the IEEE Nuclear and Plasma Sciences Society (NPSS) and the American Physical Society (APS). Li is also an honorary member of the National Academy of Inventors (NAI).

IOP Publishing

Cold Atmospheric Plasma-based Cancer Therapy
(Second Edition)

Michael Keidar, Dayun Yan and Li Lin

Chapter 1

Introduction to non-thermal plasma physics

Plasma physics is the foundation for understanding nonthermal plasma and its application. Chapter 1 briefly introduces the basic concepts in plasma physics, plasma engineering, and plasma medicine.

1.1 Plasma

Plasma is the fourth state of matter defined as an ionized gas (figure 1.1). Ionization is the process of stripping electrons from atoms and molecules that form negatively and positively charged particles [1]. Dependent on the relative density of charged particles, we can distinguish between the weakly and strongly ionized gases. A weakly ionized gas is characterized by a relatively small fraction of charged particles and its behavior can be largely described by neutral gas laws while strongly ionized gases are properly described by addition of electrodynamics coupled with gas dynamics.

A physical state of an ionized gas in which the densities of positively and negatively charged particles are approximately equal can be called a quasi-neutrality state. As such, plasma can be defined as an ionized gas satisfying quasi-neutrality conditions. The term plasma was coined by Irving Langmuir in 1920. Mott-Smith described the history of introducing the term 'plasma' in his letter to *Nature*. 'Langmuir pointed out that the 'equilibrium' part of the discharge acted as a sort of sub-stratum carrying particles of special kinds, like high-velocity electrons from thermionic filaments, molecules, and ions of gas impurities. This reminds him of the way blood plasma carries around red and white corpuscles and germs. So, he proposed to call our "uniform discharge" a "plasma"' [3].

Let us introduce the key plasma properties to help with understanding some of the terminology that will be used in this book.

doi:10.1088/978-0-7503-5537-7ch1

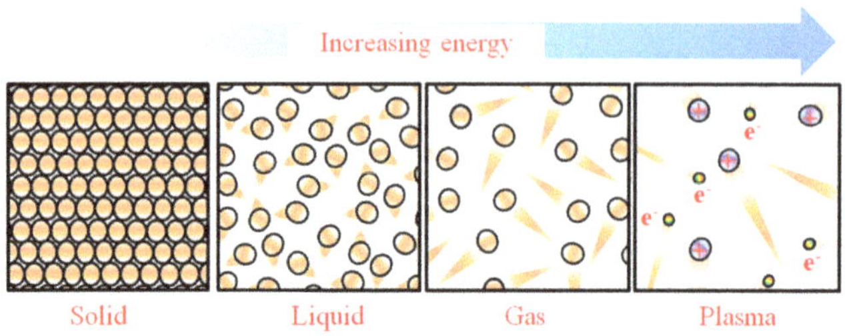

Figure 1.1. Four states of matter. Schematical illustration of solid, liquid, gas, and plasma. Brown balls, violet balls, and iridescent balls represent the neutral atoms, the positively charged ions, and electrons, respectively. Reprinted from [2], CC BY 3.0.

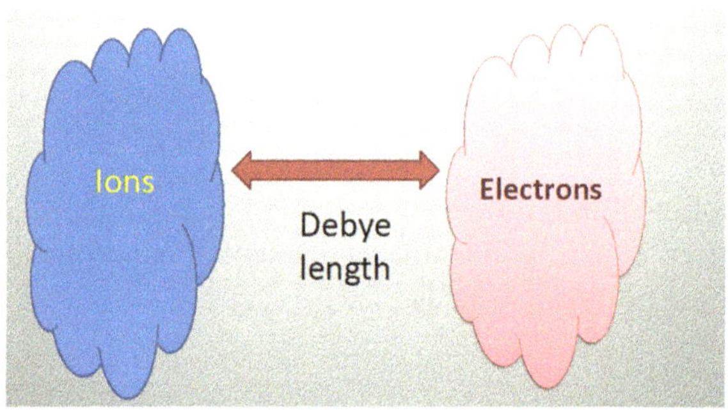

Figure 1.2. Schematics of charge separation in plasma.

1.2 Plasma quasi-neutrality

This is any perturbation in the quasi-neutral plasma such as a shift of electrons with respect to the ions, i.e., separation of charges. The charge separation produces an electric field that works to restore the unperturbed plasma state.

Let us consider an example of the quasi-neutrality phenomenon. A fully ionized plasma is formed with an electron density of about 10^{22} m^{-3} and in volume with a radius of about 1 mm. If the plasma's quasi-neutrality is violated due to charge separation at the characteristic distance of about 1 mm, an electric field of about 10^{11} V m^{-1} would appear. This means that the voltage drop of about 10^8 V will be set at a distance of about 1 mm. Such a large electric field will work to restore charge neutrality. However, if a relatively small plasma volume is considered, such an electric field will work to store the charge neutrality. However, if a relatively small plasma volume is considered, such an electric field and potential drop might not be strong enough to affect the particle motion and restore the quasi-neutrality. Thus, the quasi-neutrality condition can be violated on a small scale. The characteristics scale at which the charge separation can exist is called the Debye length (figure 1.2). The Debye length can be estimated using the following energy balance consideration.

The potential energy of the charged particle in the case of full charge separation by a distance of about L_D is the order of the particle's thermal energy $k_B T_e$, where k_B is the Boltzmann constant, and T_e is the electron temperature. On the other hand, the potential energy ($e\varphi$) due to the full separation of charges can be estimated (in the case of a planar capacitor) as $\frac{e^2 L_D^2 N_0}{\varepsilon_0}$, where e is the electron charge, N_0 is the electron density, and ε_0 is the dielectric permittivity. The energy balance yields:

$$e\varphi \sim \frac{e^2 L_D^2 N_0}{\varepsilon_0} \sim k_B T_e$$

Solving for the Debye length, one can find:

$$L_D = \sqrt{(\varepsilon_0 k_B T_e / e^2 N_0)}\,.$$

1.3 Plasma oscillations

Plasma oscillations are associated with the temporal behavior of plasma (figure 1.3). Since plasmas are electrically neutral, any violation of quasi-neutrality leads to the formation of a high electric field that leads to restoring the quasi-neutrality. Such an electric field appears due to the large difference between the ion and the electron masses. Any electron motion affects the microscale electric field, and, as a result, electrons oscillate around ions.

As the electrons move, only a net positive charge of ions will remain. It is plausible to assume that at the electron timescale, ions will not be affected since their mass is much larger. Due to charge separation, the induced electric field E will reduce this charge separation and accelerate electrons back. The electrons will gain kinetic energy, and their inertia will cause them to overshoot (pass) their original position. As a result, the plasma will become charged and separated again, and an

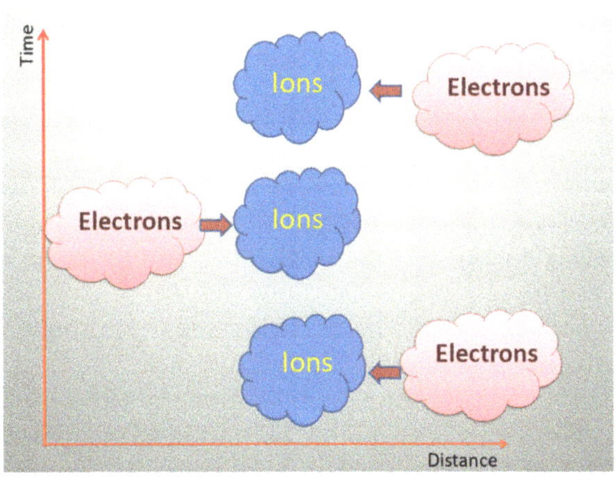

Figure1.3. Schematics electron oscillations with plasma frequency.

electric field will be formed again in the opposite direction. If there is no damping mechanism (for instance, collisions that will be discussed next), these oscillations will continue forever with the frequency:

$$\omega_p = \sqrt{\frac{e^2 N_e}{\varepsilon_0 m_e}}$$

where e is the elementary charge, N_e is the electron density, ε_0 is the vacuum permittivity, and m_e is the electron mass.

Plasma oscillations are localized and do not propagate without considering electron temperature. Thermal effects lead to the propagation of the oscillations, i.e., the generation of a plasma wave or Langmuir wave.

1.4 Particle collisions

Particles in plasma change their momentum, energy, excited state, or charge through collisions. Broadly there are two types of collisions: *elastic* and *nonelastic*. *Elastic collisions* can be defined as events in which the total kinetic energy of the particles is conserved, and the particles retain their original charges.

(1) *Nonelastic collisions* lead to the kinetic energy redistribution between the particles and transfer kinetic energy into the internal mode of one or more colliding particles. Nonelastic collisions lead to direct and reverse processes, which include particle excitation and deexcitation, ionization and recombination, charge-exchange, dissociation, and charge neutralization. Among nonelastic collisions, ionization plays a very important role in sustaining the plasma. A number of ionization pathways can be distinguished in the typical plasma system:

(2) *Direct ionization* by electron impact is the ionization of neutrals from the ground state by electrons having a high enough energy to cause ionization in a single act. The reverse process is the recombination by which ions capture the free electrons to neutralize the ion into an atom. When an ion captures two electrons, the process is called three-body recombination, and if it captures one electron, it is called radiative recombination.

- *Stepwise ionization* by electron impact is the ionization of a previously excited neutral species.
- *Ionization of collisions of heavy atoms.* This is the ionization in which the particles' kinetic energy can lead to the ionization process. If chemical energy is important, such a process is called associative ionization.
- *The photoionization process.* This is the process in which the ionization is caused by the photons' interaction with neutral atoms. The reverse process is called photorecombination. When a molecular gas is ionized, the reaction can go into excitation of the molecule as well as into the molecule's dissociation. Dissociation is the process when a molecule decays into separate atoms.

- *Charge-exchange collisions.* Positive ions can collide with atoms and capture a valence electron, resulting in charge transfer from the atom to the ion.
- *Associative ionization* is a gas-phase chemical reaction in which two atoms/molecules collide and form an ion via interaction. The energy that is released as a result of the chemical reaction is transferred into the electron's energy. Ionization occurs if this energy exceeds the ionization potential. One or both colliding species may be in an excited state.
- *Penning ionization* involves a reaction between neutral atoms/molecules. This process is named after the Dutch physicist Frans Michel Penning (effect reported in 1927). The term Penning ionization refers to the interaction between an excited-state atom/molecule and a target atom/molecule.
- *Coulomb collisions* are elastic collisions between charge particles, such as electron–electron, electron–ion, and ion–ion collisions.

1.5 Plasma types

Our daily life involves plasma in various ways. The spark that jumps between two contacts represents the plasma associated with an electrical spark discharge in air, plasma displays have captured the industry's attention due to the quality of the picture they can produce. Nevertheless, under ordinary conditions on Earth, plasma is rather a rare phenomenon, but in the Universe, cold solid bodies are an exception. Most of the matter in the Universe is ionized and thus in the plasma state (about 99.9%). Various mechanisms produce plasma in the Universe. In the stars, the neutral atoms are ionized due to high temperatures. Interstellar gases are ionized due to the ultraviolet radiation from the stars.

Plasmas produced in Nature and laboratory plasmas are characterized by a wide range of temperatures and pressures. Using the Debye length and the plasma frequency, which depend on the electron number density and electron temperature, it is possible to classify plasmas as rarified, dense, classical, and quantum. They span from high-temperature plasmas in thermonuclear fusion reactors to room-temperature cold plasmas in some biomedical applications [4, 5]. Figure 1.4 displays the diagram of various natural and laboratory plasmas with their typical plasma density and electron temperatures. This book focuses exclusively on cold plasma's application in bioengineering, particularly cancer therapy [6].

1.6 Cold atmospheric plasmas (CAP)

Recent progress in atmospheric plasmas led to the production of cold plasmas with ion temperatures close to room temperature. To distinguish from traditional thermal plasma, cold plasmas are also named nonthermal plasma. Cold plasma has a very low-level ionization rate. Though electron temperature is still relatively high, the heavy particles (ions, neutral atoms) have much cooler temperature, which is due to weak elastic collisions in a non-equilibrium discharge process (figure 1.5).

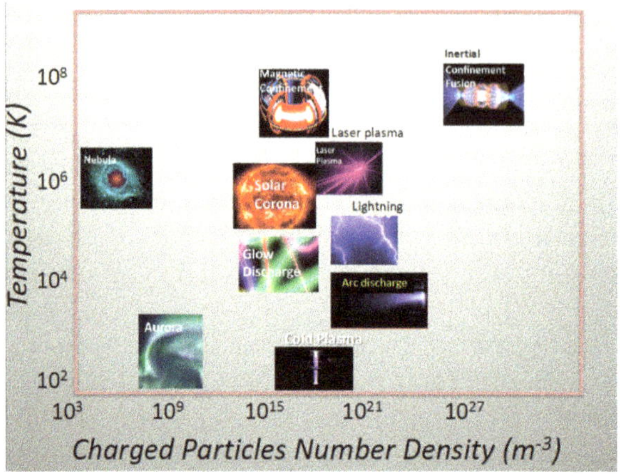

Figure 1.4. Diagram showing various natural and laboratory plasmas.

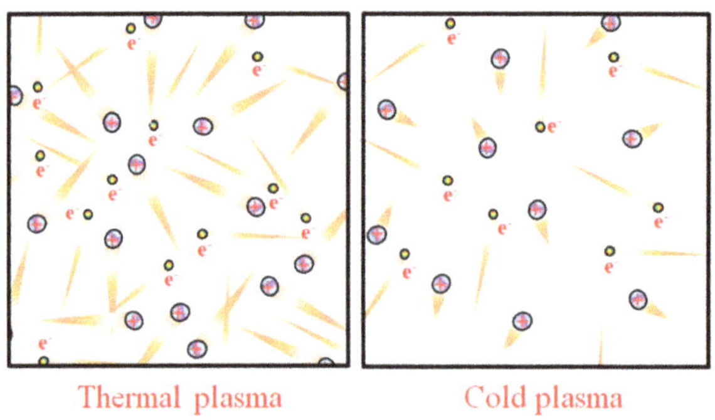

Figure 1.5. Thermal plasma and non-thermal plasma. Only ionized particles were drawn here. Brown balls, violet balls, and iridescent balls represent the neutral atoms, the positively charged ions, and electrons, respectively. Reprinted from [2], CC BY 3.0.

The features make cold plasma even different from typical low-temperature plasma of a typical electrical discharge. Such plasmas constitute very useful tools enabling interaction with biological tissue without thermal damage. Cold plasma technology development resulted in rapidly forming a new field of plasma medicine [7, 8].

In general, the CAP sources can be classified into two major groups according to the principal mechanism of generation and application [9]. Schematically two approaches are shown in figure 1.6.

(a) *Direct plasma sources* utilize living tissue or organs as one of the electrodes, and thus, living tissue directly participates in the active discharge plasma processes. Some current may flow through living tissue in the form of small conduction current, displacement current, or both. To this end conduction current must be limited to avoid any thermal effects or electrical stimulation of the muscles. The dielectric barrier discharge (DBD) is a typical example of direct plasma sources.

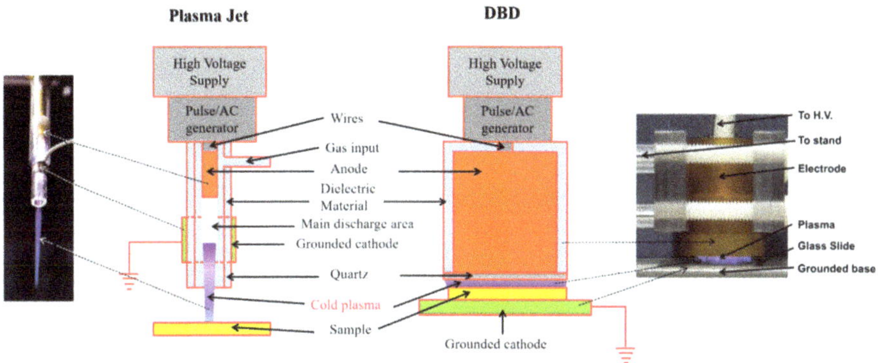

Figure 1.6. Atmospheric pressure plasma jet (APPJ) and dielectric barrier discharge (DBD) are the main CAP devices used in plasma medicine. The same components in APPJ (direct discharge) and DBD (indirect discharge) are drawn with the same colors. Left image reprinted from [14], CC BY 4.0. Middle image reprinted from [2], CC BY 3.0. Right image reprinted from [15], CC BY 4.0.

(b) *Indirect plasma sources* are produced between two electrodes and are then transported to the area of application entrained in a gas flow (typically helium or argon). A great variety of configurations of indirect plasma sources exist in size, type of gas, and power. They range from very narrow 'plasma needles', atmospheric pressure plasma jets (APPJs) to larger 'plasma torches'. Corona discharge plasmas have similar properties to those of APPJs.

While the effects of an electric field on cells have been known for a long time, CAP produces a much larger array of possible pathways due to the presence of various chemically active species and charged particles in addition to the electric field. CAP interaction with cells leads to various effects, including cell detachment without affecting cell viability, controllable cell death, necrosis, cell migration etc. The aim of using CAP in biomedical applications can be mainly achieved by operating under the thermal damage threshold and inducing a chemically specific response or modification. The chemical reaction can be promoted by tuning the pressure, gas composition, and discharge power.

CAP interaction with cells and tissue can elicit multiple pathways. The variety of different effects of plasma can be explained by their complex chemical composition and variations in the way that CAP is generated. CAP is a cocktail containing a variety of reactive oxygen species (ROS), reactive nitrogen species (RNS), charge particles, electric field, UV, etc (figure 1.7).

1.7 CAP in plasma therapy

This book focuses mainly on cancer therapy that utilizes CAP as a new modality for cancer treatment [10]. A high-level understanding of CAP interaction with cells and tissue consists of the notion that chemical elements produced by CAP in gaseous and liquid phases are potentially toxic. This includes ROS, which might promote a

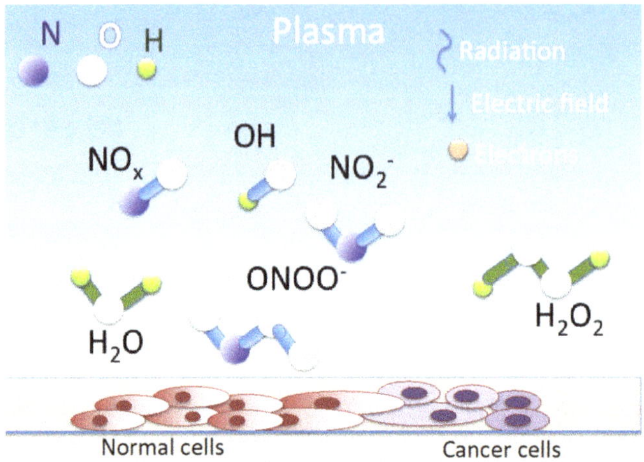

Figure 1.7. Plasma interfaces the living cells. Components of cold atmospheric plasma. Reprinted from [11], copyright (2018), with permission from Elsevier.

'plasma killing effect,' while others, such as RNS could produce a 'plasma healing' effect. Forming various combinations of these species might provide great potential for activating specific signaling pathways in cells. CAP treatment possesses powerful lethal capabilities against tumor cells both *in vitro* and *in vivo*, and just as importantly, the normal counterpart cells have been shown to be less sensitive to the same CAP treatment. All effects mentioned earlier are related to plasma chemistry. At the same time, the role of other plasma effects, such as charged particles and electric field, remains elusive. In this book, we will describe some recent advances in understanding the physical effects of the electromagnetic (EM) field.

One of the unique features of plasmas compared to other sources of reactivity is the ability to very rapidly (within ns) change the reactive species production pathways, thereby enabling feedback systems that customize *in real time* the reactivity delivered to cells [11]. At one extreme, this enables personalized medicine, and at the other extreme can address cell and tissue variability while treating a single patient as an example. That is, plasma can be engineered to affect specific cellular or tissue functions.

Schematically the adaptive CAP is shown in figure 1.8. Practical implementation of adaptive CAP requires real-time monitoring of the plasma-living tissue interaction leading to the adaptation of plasma parameters based on feedback control. The related idea of plasma device feedback control and managed delivery of plasma dose was described elsewhere [12]. An example of feedback-based adaptive plasma utilizing a RealTime™ assay provided evidence that cellular response to plasma treatment can be monitored in real time [13]. In this book we will describe recent advances in adaptive plasmas, including the development of an intelligent plasma capable of making decisions without an external computer.

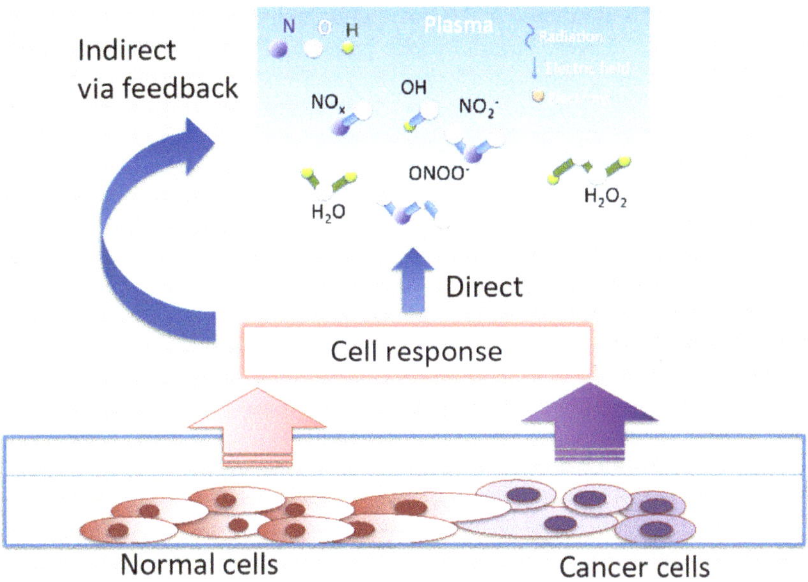

Figure 1.8. Adaptive cold atmospheric plasma concept based on differential treatment of cancer and normal cells. The feedback system modulates the chemical composition of plasma, leading to adaptation. Direct feedback or plasma self-adaptation is based on self-organization and pattern formation. Reprinted from [11], copyright (2018), with permission from Elsevier.

References

[1] Keidar M and Beilis I I 2018 *Plasma Engineering* 2nd edn (Oxford: Elsevier-Academic Press)

[2] Yan D, Sherman J H and Keidar M 2016 Cold atmospheric plasma, a novel promising anti-cancer treatment modality *Oncotarget* **8** 15977

[3] Mott-Smith H M 1971 History of 'plasmas' *Nature* **233** 219

[4] Fridman G, Friedman G, Gutsol A, Shekhter A B, Vasilets V N and Fridman A 2008 Applied plasma medicine *Plasma Processes Polym.* **5** 503–33

[5] Laroussi M, Lu X and Keidar M 2017 Perspective: the physics, diagnostics, and applications of atmospheric pressure low temperature plasma sources used in plasma medicine *J. Appl. Phys.* **122** 020901

[6] Keidar M 2015 Plasma for cancer treatment *Plasma Sources Sci. Technol.* **24** 033001

[7] Kong M G, Kroesen G, Morfill G, Nosenko T, Shimizu T, Van Dijk J and Zimmermann J L 2009 Plasma medicine: an introductory review *New J. Phys.* **11** 115012

[8] Fridman A and Friedman G 2013 *Plasma Medicine* (London: Wiley)

[9] Weltmann K D, Kindel E, von Woedtke T, Hähnel M, Stieber M and Brandenburg R 2010 Atmospheric-pressure plasma sources: prospective tools for plasma medicine *Pure Appl. Chem.* **82** 1223–37

[10] Keidar M, Shashurin A, Volotskova O, Ann Stepp M, Srinivasan P, Sandler A and Trink B 2013 Cold atmospheric plasma in cancer therapy *Phys. Plasma* **20** 057101

[11] Keidar M, Yan D, Beilis I I, Trink B and Sherman J H 2018 Plasmas for treating cancer: opportunities for adaptive and self-adaptive approaches *Trends Biotechnol.* **36** 586–93

[12] Gidon D, Abbas H S, Bonzanini A D, Graves D B, Mohammadpour Velni J and Mesbah A 2017 Effective dose delivery in atmospheric pressure plasma jets for plasma medicine: a model predictive control approach *Plasma Sources Sci. Technol.* **26** 085005

[13] Gjika E, Pal-Ghosh S, Tang A, Kirschner M, Tadvalkar G, Canady J, Stepp M A and Keidar M 2018 Adaptation of operational parameters of cold atmospheric plasma for *in vitro* treatment of cancer cells *ACS Appl. Mater. Interfaces* **10** 9269–79

[14] Siu A, Volotskova O, Cheng X, Khalsa S S, Bian K, Murad F, Keidar M and Sherman J H 2015 Differential effects of cold atmospheric plasma in the treatment of malignant glioma *PLoS ONE* **10** e0126313

[15] Kalghatgi S, Kelly C M, Cerchar E, Torabi B, Alekseev O, Fridman A, Friedman G and Azizkhan-Clifford J 2011 Effects of non-thermal plasma on mmmalian cells *PLoS ONE* **6** e16270

IOP Publishing

Cold Atmospheric Plasma-based Cancer Therapy
(Second Edition)

Michael Keidar, Dayun Yan and Li Lin

Chapter 2

Cold atmospheric plasma generators

Chapter 2 summarizes the methods to generate plasmas with examples of cold atmospheric plasma (CAP) generators. This includes the concept of cold atmospheric plasma, the detailed physics behind the ionizations, and the hardware design for making such ionizations. The latest inventions, such as plasma jet array and micro-jet, are also introduced in this chapter, including their advantages and limits.

2.1 Ionization

In chapter 1, we introduced that plasma was a fully or partially ionized gaseous fluid. Plasma needs enough energy input to maintain the ionization and thus maintain its densities of free electrons and ions. There are many ways to apply energy on gases to make plasmas. In addition to heating the gas, the most common way to ionize the gases and create plasmas is to input electromagnetic waves. Of course, electromagnetic wave irradiation can also heat the gas. Figure 2.1 summarizes the types of ionizations along with the input electromagnetic field frequencies. From DC (no frequency) to the kHz range, the most common plasma generator design applies high voltage on a pair of electrodes with gas between, namely capacitively coupled plasma (CCP). The voltage will thus ionize the gas. However, when the frequency is higher such as in the MHz range, the most common generator design is to apply a current source on a coil, namely inductively coupled plasma (ICP). The MHz current in the coil will generate an oscillating electromagnetic field in the gas to ionize. ICP is usually working at a higher frequency than CCP because the impedance of a coil can be higher at a higher frequency. In comparison, the impedance of a capacitor is higher at a lower frequency. The plasma generator is a load in the circuit, and one needs a significant impedance of the load to apply power to it; otherwise, the circuit will be hot with high energy waste and even shorted. When the frequency is in the microwave range or higher, the incoming

doi:10.1088/978-0-7503-5537-7ch2

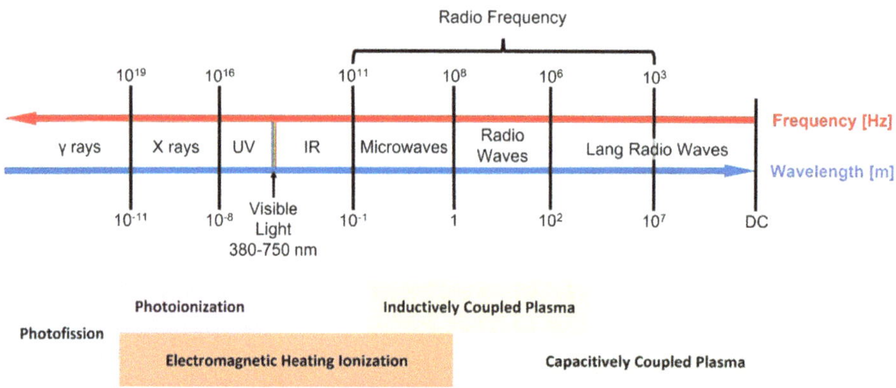

Figure 2.1. Electromagnetic wave ionizations.

electromagnetic wave becomes a beam, and the heating effect becomes significant. When the beam is ultraviolet (UV) or with a higher frequency, single-photon energy can be higher than the ionization threshold of molecules to make photoionization. For the γ rays, however, even photofission becomes possible, which can knock out electrons from the molecules and dissociate the nucleus. These ionization methods will be introduced in detail in the following sections.

2.1.1 Heating and laser ionization

Heating is the most straightforward method of creating plasma. When a gas is heated to a sufficiently high temperature, its molecules have thermal velocities that are high enough to ionize others during particle collisions. The gas thus becomes plasma. This process is called thermal ionization. The temperature required for thermal ionization depends on the gas being ionized and the pressure it is under. Laser heating is a process that uses a high-energy laser beam to transfer energy to a gas and increase its temperature. The gas molecules absorb the laser beam, which then vibrates and rotates faster, increasing their thermal velocities during collisions. When the gas reaches a specific temperature, the molecules become ionized, and the gas becomes plasma.

Laser photoionization is another story. The laser beam is tuned to a specific frequency, which is higher than the ionization energy of the gas molecule. Usually, it is at least in the UV range (see figure 2.1). When the laser beam interacts with the gas molecule, it transfers its energy to the electron in the molecule, which is then ejected from the molecule, creating a positive ion. Different atoms have different ionization thresholds. The electron farthest from the atom nucleus is with the weakest force from the nucleus and thus is the first electron to lose. The energy to knock it out is the first ionization threshold. The second farthest one requires the second ionization threshold much higher than the first one. Therefore, in most cases, +1 ion density is usually much higher than +2 ions.

2.1.2 Microwave ionization and inductively coupled plasma

Microwave ionization is another process that uses electromagnetic radiation to ionize a gas. There are two types of microwave ionization: microwave beam irradiation and ICP. The former can heat the gas and is more like the previously mentioned laser beam irradiation, while the latter generates an oscillating electromagnetic field in the gas to ionize. To create an ICP, a coil of wire is wrapped around a quartz tube or a vessel with other materials that are transparent to the specific frequency of electromagnetic waves produced by the coil [1, 2]. The vessel contains the gas to be ionized. An alternating current is then passed through the coil, which creates a rapidly changing magnetic field that induces an electric field in the gas. A neutral gas has a low density of free electrons due to the cosmos ray and quantum uncertainty. These electrons are thus known as 'seed electrons' [3]. The electric field accelerates the seed electrons charged with particles (such as electrons and ions) in the plasma, causing them to collide with other gas molecules. These collisions can lead to ionization and excitation of the gas molecules, generating a cascade of ionization events that rapidly increase the plasma density. In other words, more electrons will thus ionize and lead to more electrons, known as an 'avalanche'. Plasma is thus generated.

One of the advantages of microwave ionization is that microwaves are transmitted through a waveguide, and plasma can be generated over a large volume of gas, making it useful for processing large batches of material. Depending on the gas used and the operating conditions, microwave plasma can also generate a wide range of plasma chemistries. For example, using a mixture of gases can create plasmas that contain specific chemical species, such as nitrogen or oxygen. This can be useful for surface treatments, where plasma chemistry can be tailored to achieve specific surface properties. In addition to its use in industrial applications, microwave ionization is used in research settings to study plasma physics and chemistry. By controlling the operating conditions and the gas composition, researchers can study the behavior of plasmas and their interactions with materials.

However, a unique feature of ICP is its magnetic field. The ICP also causes the charged particles in the gas to move in a circular or helical path, depending on the field's orientation [4]. This movement is known as cyclotron motion or Larmor motion. When an alternating magnetic field is applied, the direction of the magnetic field changes rapidly, which causes the charged particles in the gas to undergo a complex motion known as $E \times B$ drift [4]. In this process, the electric field induced by the magnetic field causes the charged particles to move in a circular path around the axis of the magnetic field. This circular motion causes the charged particles to collide with each other, transferring energy and increasing the temperature of the gas. Recent research includes adding one or more external magnetic fields at DC or lower frequency. The matching between these two fields thus makes more interesting but complicated behaviors and industrial applications.

The microwave beam and ICP can create plasma over a large volume of gas, making it useful for processing large batches of material and at relatively low pressures if required. Depending on the gas used and the operating conditions, they can also generate a wide range of plasma chemistries.

2.1.3 Capacitively coupled plasma

Capacitively coupled ionization is a process that uses an alternating electric field to ionize a gas. Such plasma is known as CCP [5–7]. In this method, two metal plates are placed near each other, with gas between them. Therefore, CCP is very commonly seen in nature and daily life, such as lightning, the spark between two high-voltage wires when closing in, and plasma balls. To create CCP, a DC or AC (usually kHz) voltage is applied to the plates, giving us an electric field between them. When the field is strong enough, a complicated physical process, namely 'streamer propagation' will occur (figure 2.2) [8]. In the vicinity of one of the electrodes, when it has a high enough positive potential, the electric field is strong enough to accelerate the seed electrons. The electrode is considered an anode at this moment. During massive random seed electron collisions, avalanche ionizations occur. However, those electrons with energies lower than the ionization thresholds cannot ionize but may excite the molecules. Those excited molecules emit photons, making the plasma visible [9, 10]. The avalanche process is much faster than the kHz AC external field oscillation. Therefore, the external electric field is a DC w.r.t. the electrons. The electrons share the same directional drift velocity toward the anode. After the electrons leave their original location, positive ions remain due to their heavy weights and let the original location positively charge. This region is named 'streamer head', performing as a new

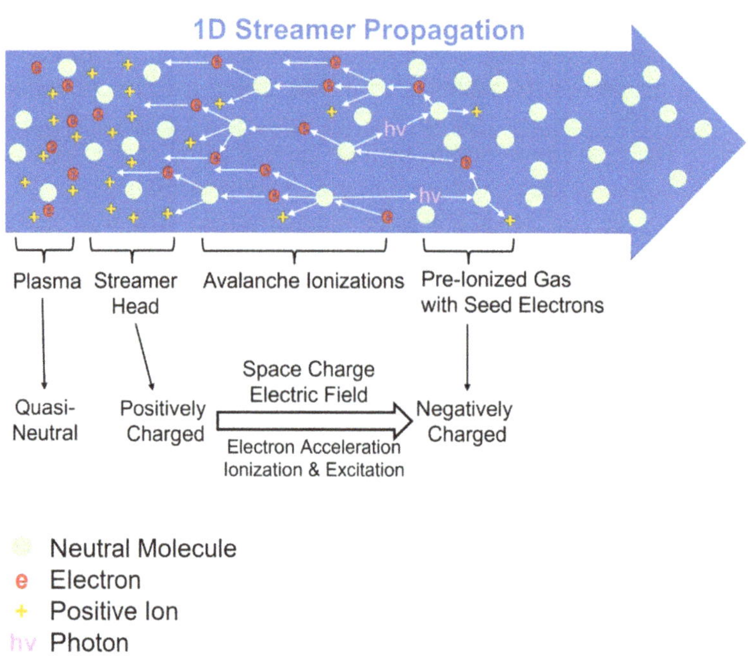

Figure 2.2. Streamer propagation.

anode. More seed electrons nearby will be attracted, and an avalanche (making more electrons) will be on its way, moving to the streamer's head. Finally, these electrons will neutralize the streamer head, but a new streamer head appears because these electrons left their original location. Overall, the above process repeats, and a high-ionization-rate region moves toward the cathode with strong photon emissions while the electrons are moving oppositely, and no ions need to move. Therefore, such a streamer propagation is also considered a kind of ionization wave, a pulse waveform of high ionization rate propagation [11–13]. When the streamer head, in other words, the ionization wavefront, reaches the cathode, a plasma path is created between the electrodes with a significantly higher ionization degree. There are many more further analyses about such a CCP start-up process. Specific generators will be introduced in the next section of this chapter.

After such a start-up process, an AC-powered CCP will continue to repeat such a process in each discharge period. Before a later streamer appears, the electrons and ions may recombine in the plasma. Still, a certain level of electron density will remain, and the next streamer will propagate following such a weak plasma track with more of those seed electrons. This means that a later streamer can make a higher electron density due to more seed electrons being involved in the electron-impact ionizations. However, more electrons also mean a higher recombination rate, and the input electric field energy needs to be shared with more electrons. Therefore, the increment of electron density along the repetition of streamer propagation will finally stop when the increment of ionization rate is balanced by the increment of recombination rate and the division of electric field (also known as the reduced electric field, denoted as E/N with [Td] as the unit, where E is the electric field, and N is the particle number). At this moment, the AC-powered CCP is in a steady state.

For a DC-power CCP, however, it only has a one-time ionization wave to start up, before reaching a steady state. The ignition condition mainly depends on the pressure, gap distance between the electrodes, and the applied voltage, known as *Paschen*'s law. A classic model of DC-power CCP is a DC discharge tube. Significant striations (the presence of bright and dark regions along the length of the tube) will finally be established at the steady state in the tube when the discharge voltage is low, and the status is considered as a glow discharge. When the applied voltage is increased, the ionization degree is increased (more molecules are ionized), the total impedance of the plasma is decreased, and thus, the current is boosted up. The system will finally transit to the arc. This is a typical plasma lamp schematic. Please note that the actual process is more complicated, such as the mechanism of generating the striations, but the model is a very classic CCP found in almost all plasma textbooks [3, 14]. However, the traditional discharge tube model usually works at low pressure rather than atmospheric. Therefore, we will not go deeper into this topic.

2.2 Cold atmospheric plasma

Cold plasma or low-temperature plasma is a thermal non-equilibrium plasma, meaning that the electron temperature is high (usually with an average value above 3 eV, or higher than 34 800 K) and a low gas bulk temperature, in other words, the macroscopic gas temperature usually close to the room temperature. CAP is thus a temperature with atmospheric pressure. Such a difference between electrons and other particles is because of the low mass of electrons, making them the first type of charged particle to be accelerated by the external electric field. The electron drift velocity will be converted into random thermal velocity during collisions. Meanwhile, during collisions, the momentum of electrons will be transferred to other heavier particles and, thus, heat the macroscopic gas. Statistically, the momentum transfer will finally stop when the thermal equilibrium is reached, but it takes time. Therefore, one can have cold plasma by applying relatively lower power on gas to ionize. However, ensure the system contains a mechanism that can stop the momentum transfer before the gas bulk temperature is too high. For example, with O_2 added, free electrons can be captured by the attachment reaction $e + O_2 \Rightarrow O^- + O$. By losing electrons, the momentum transfer is limited, and the system can be balanced at a state of high electron temperature but low gas bulk temperature.

CAP is hard to generate by ionizing polyatomic molecules with chemical bonds to absorb energies and therefore requires higher energy input to ionize than monoatomic molecules. In other words, the chemical bonds can store energy and heat up (such as by vibrational and rotational excitation, see section 2.1.1) the gas before ionizing. Once the polyatomic molecules are ionized, the gas usually has a much higher bulk temperature than the monoatomic gases. Therefore, to generate CAP, a noble gas (also known as inert gas and rare gas) such as helium with monoatomic molecules is needed, while ionizing air will result in a much higher gas bulk temperature.

2.2.1 CAP jet

CAP is usually generated by a CCP setup in the open air. With the cathode designed as a hollow structure, such as a ring, the streamer can pass through it and reach a further distance, namely the plasma jet. As shown in figure 2.3, the most used designs of CAP jet generators are in-ring and ring–ring setups [15, 16]. The only difference is the design of the anode. Both designs can generate an axial electric field towards the nozzle exit. Therefore, a streamer (see section 2.1.3) can propagate through the inert gas flow field from the anode to the exit. An example of such a helium CAP jet with a pin–ring setup is shown in figure 2.4. Please note that the tip of this CAP jet sometimes may be bent upwards by the buoyancy force of the helium in ambient air, depending on the gas flow rate. The streamer propagation can also be observed and analyzed through an intensified charge-coupled device (ICCD) high-speed camera which is synchronized with the plasma generator [10, 17]. In figure 2.5, the exposure width is 10 ns. However, if the exposure width is extended, the temporal resolution will be decreased, and the plasma bullet image will be longer, as shown in figure 2.6, where the exposure width is 100 ns. This is equivalent to the overlap of multiple brightness distributions of 10 ns bullets. Please note that a commonly seen corona discharge usually occurs in ambient air and at a high-voltage

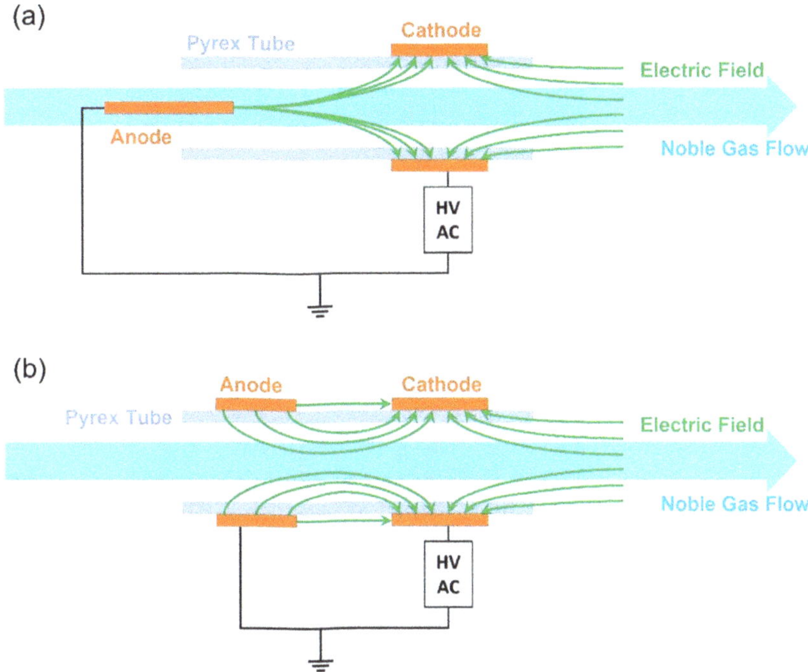

Figure 2.3. Two most common designs of CAP jet generators. (a) Pin–ring electrodes setup. (b) Ring–ring electrodes setup.

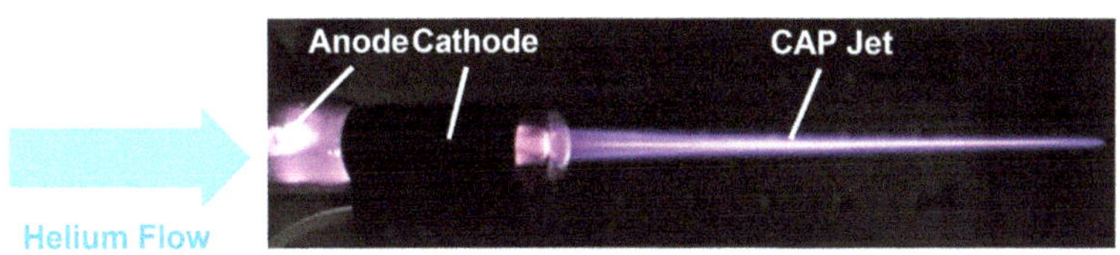

Figure 2.4. A photo of a helium CAP jet.

electrode with sharp edges. Such a geometry leads to a high potential gradient, such as a strong electric field, to generate plasma. This feature can always be found in discharges at the tip of a pin.

2.2.2 CAP jet array

An upgraded version of a plasma jet is a jet array that can be a 1D or 2D cluster of plasma jet generators, as shown in figure 2.7. An array setup can cover a much larger treatment area than a single jet [18]. However, there are also unsolved technique issues in the plasma jet array. One of the main challenges is ensuring that all the individual plasma jets are identical and have similar characteristics. This can be achieved by carefully controlling the gas flow, voltage, and other parameters that

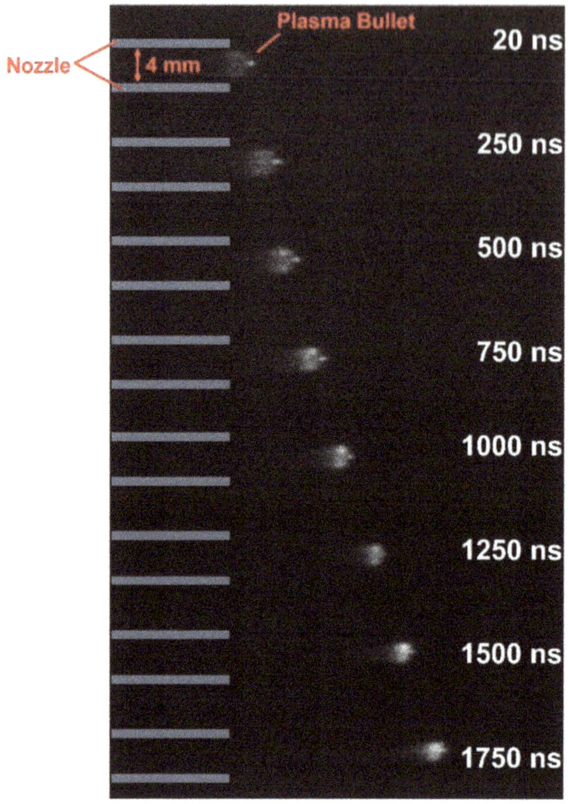

Figure 2.5. The streamer propagation (plasma bullet) of a helium CAP jet captured by a synchronized ICCD high-speed camera and the text in white showing the time axis.

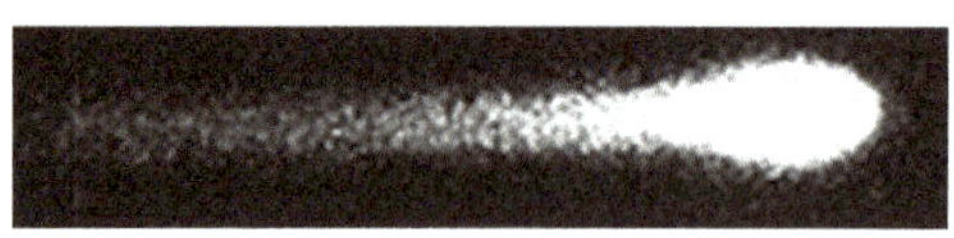

Figure 2.6. A plasma bullet image captured with an exposure time of 100 ns.

affect plasma formation. For example, in figure 2.7, there are multiple flow splitters to divide the input gas flow; otherwise, more input gas flow rate will be delivered to the middle nozzle. However, internal flow control is still an unsolved and not optimized problem that requires future research. The external flow field is also a current research topic for multi-jet setups. Many publications have shown that the CAP jet at the edge of the array suffers a spread of gas flow field [19–21]. Also, the electric forces between the positively charged streamers may expel the plasma jet from each other. Thus, there can be a tendency for the jets to spread out and become less uniform, especially at higher power levels. But in certain cases, they can also merge into one jet, such as when the seed electron clouds are merged due to a small

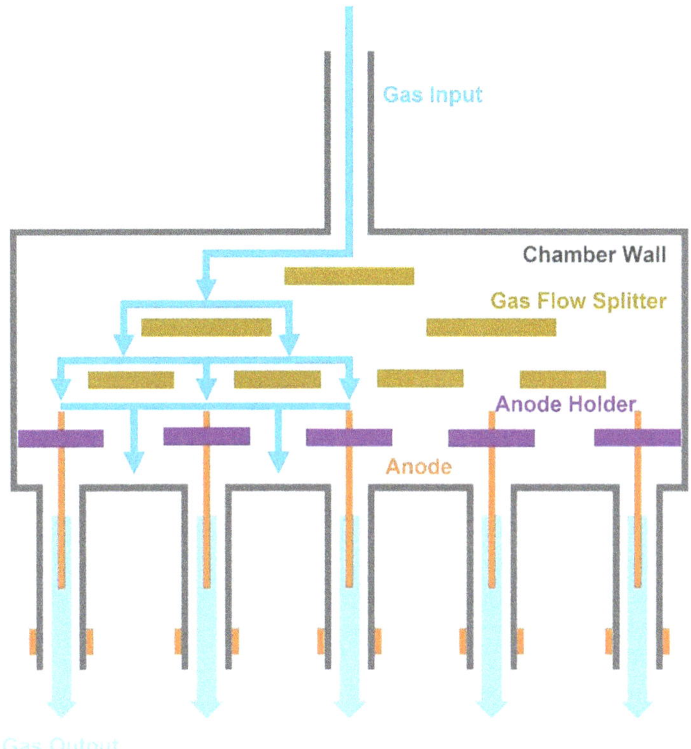

Figure 2.7. A multiple CAP jet design. Please note that the gas flow path is marked in light blue and the internal path should be symmetric in an ideal case. Only a half of the flow path on the left is shown in the figure.

gap among the nozzles [22, 23]. This can be mitigated using special electrode designs or adjusting the gas flow and other parameters to control the jet shape and stability. In addition, if the jets are too far apart, there may be areas of untreated material between the jets. Moreover, a plasma jet array needs much more power and gas input. For example, the jets in a 5 by 5 array share the same discharge voltage, but the current will be about 25 times higher than the gas flow rate. Overall, the multi-jet design is currently at the frontier of the CAP community and still needs substantial work to optimize.

2.2.3 Microscopic CAP jet

Another example of a practical design of a plasma jet is the micro-plasma jet or μCAP [24–28]. The μCAP is the opposite of the plasma jet array, designed to achieve a tiny but accurate treatment pinpoint. This is a typical design of minimally invasive surgery (MIS), a surgical technique that uses specialized tools and instruments to perform surgical procedures with minimal damage to surrounding tissues. Compared with regular surgery, the MIS achieved by using μCAP can have several advantages: (1) The small incisions used in μCAP-MIS cause less tissue damage and result in smaller scars, leading to reduced pain and a faster recovery time;

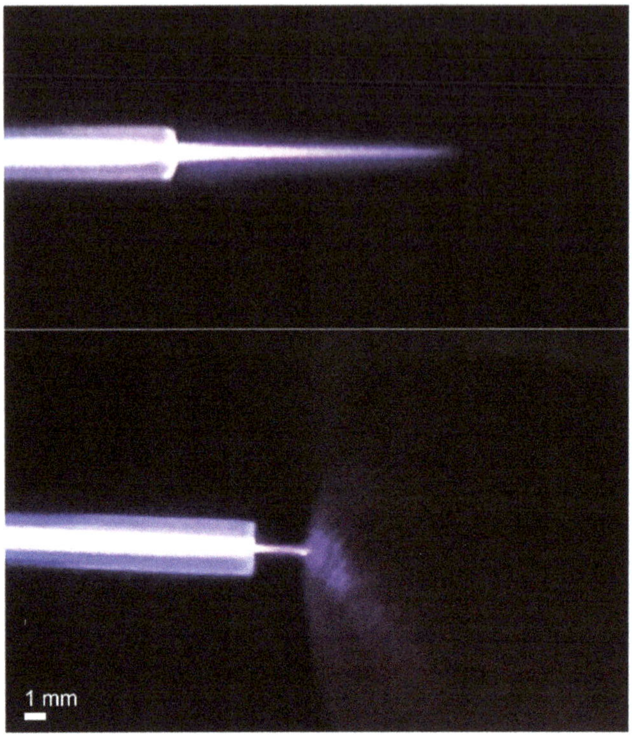

Figure 2.8. An example of μCAP.

(2) the μCAP-MIS is associated with a lower risk of complications, such as infections and bleeding, due to the smaller incisions and reduced tissue damage; (3) the μCAP-MIS typically requires a shorter hospital stay than traditional surgery, as the recovery time is faster.

An example of μCAP is shown in figure 2.8. The generator is like a regular CAP jet, and the only different part is the nozzle and a lower helium flow rate of about 0.1 LPM. The outer diameter of the nozzle is about 3 mm, and the inner diameter, which determines the μCAP diameter, is less than 1 mm. There are several *in vivo* experiments published where mice brain tumors were treated.

2.2.4 CAP discharge tube

CAP discharge tube is a newly raised CAP generator. There are two types of capacitive coupling cold atmospheric plasma designs: axial and radial discharge tubes (figure 2.9) [29, 30]. Both are designed to treat specific areas with minimal impact on surrounding tissues. The axial discharge tube is like a regular cold atmospheric plasma jet but with the nozzle sealed. Instead of a gas supply, it contains a helium–air mixture at atmospheric pressure within an enclosed tube-like chamber. The same circuit can power this type of tube as a plasma jet. On the other hand, the novel radial discharge tube contains a central rod electrode and copper

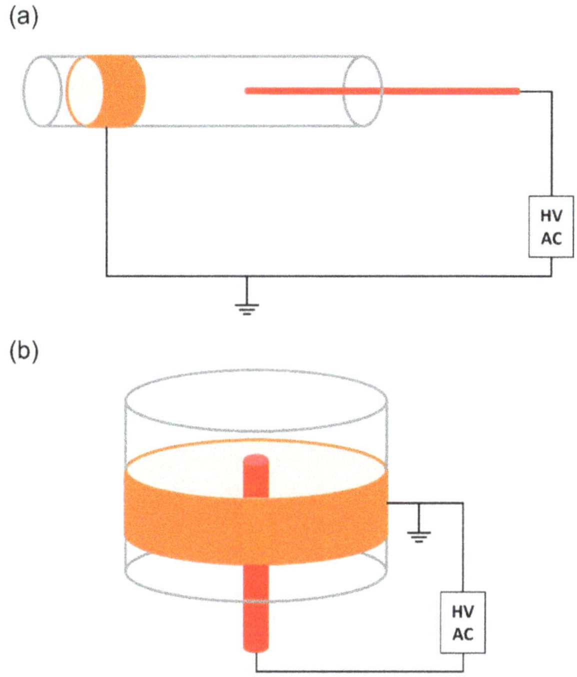

Figure 2.9. Typical designs of discharge tubes. (a) The axial CAP discharge tube. (b) The radial CAP discharge tube.

tape wrapped around the outside of the tube. This generates a radial discharge, like a plasma ball, covering a larger treatment area than the axial discharge tube. Both discharge tubes have the advantage of isolating the plasma chemistry and allowing only physical emissions to treat the target area, such as cancer cells in biomedical applications. This allows for further research into the physical effects of plasma, such as radiofrequency emissions that may resonate with the cell membrane and cause local membrane damage and sensitization. In summary, the two types of capacitive coupling cold atmospheric plasma discharge tubes offer a new approach to minimally invasive surgery and other biomedical applications. The ability to isolate plasma chemistry and treat specific areas with minimal impact on surrounding tissues makes them a promising area of research for the future.

The use of cold atmospheric plasma in biomedical applications is a rapidly developing field, with potential applications ranging from cancer treatment to wound healing. The design of discharge tubes, such as the axial and radial discharge tubes, offers a new approach to delivering plasma to specific areas with minimal impact on surrounding tissues. One potential advantage of using discharge tubes is that they can isolate the plasma chemistry and allow only physical emissions to treat the target area. This means that researchers can study the physical effects of plasma, such as radiofrequency emissions, and their potential impact on different types of cells. The physical effects will be discussed in later chapters.

2.2.5 Dielectric barrier discharge

Dielectric barrier discharge (DBD) is an electrical discharge that occurs when a high voltage is applied across two electrodes separated by a dielectric material, such as glass, ceramic, or gas layer. In a typical DBD setup, one electrode is connected to a high-voltage power supply while the other is grounded (figure 2.10) [31, 32]. When the voltage is applied, plasma is generated between the two electrodes, characterized by a bright glow and emitting UV radiation.

One of the key features of DBD is that it occurs at atmospheric pressure, making it a useful tool for various applications. For example, DBD can be used for surface treatment and modification, such as improving the adhesion of coatings to plastic surfaces or enhancing the hydrophilicity of textiles. DBD can also be used for air purification, as the UV radiation emitted by the plasma can break down harmful compounds, such as volatile organic compounds and bacteria, into harmless byproducts. This makes DBD a potentially useful technology for improving indoor air quality and reducing the spread of airborne diseases. Another potential application of DBD is in plasma medicine, where it is being investigated as a possible treatment for various diseases and conditions. One promising area of research is the use of DBD for wound healing. DBD has been shown to stimulate the production of growth factors and other signaling molecules that can promote tissue regeneration and healing. This makes it a potentially useful technology for treating chronic wounds, such as diabetic foot ulcers, that are difficult to heal using traditional approaches. DBD is also being investigated as a possible treatment for

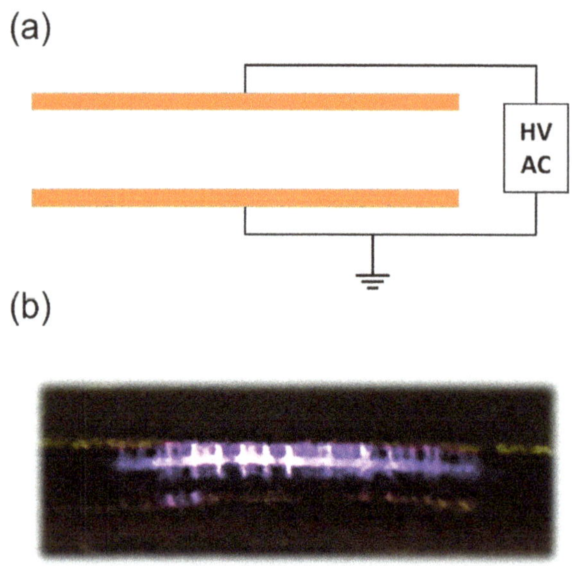

Figure 2.10. Dielectric barrier discharge. (a) Typical DBD generator design. (b) An atmospheric DBD example in ambient air without any inert gas input. The gap is 1 mm with an 18 kHz and 8 kV sinusoidal discharge voltage.

cancer. One approach involves using DBD to generate reactive oxygen and nitrogen species, which can induce apoptosis (programmed cell death) in cancer cells. DBD has been shown to be effective against various cancer cell lines *in vitro*, and there is ongoing research to explore its potential as a cancer therapy. Another area of research is the use of DBD for tissue engineering. DBD can be used to modify the surface properties of biomaterials, such as scaffolds, to promote cell adhesion and proliferation. This makes it a potentially useful technology for creating engineered tissues and organs.

DBD has several advantages and disadvantages compared to cold plasma technologies like plasma jets. The benefits of DBD include: (1) DBD can cover a relatively large area, making it potentially helpful in treating larger tissue surfaces; (2) DBD typically uses lower gas flow rates than plasma jets, which can make it easier to control the plasma environment, with less helium consumption, and reduce the risk of gas exposure for patients and clinicians; (3) DBD can be easily scaled up or down to fit different situations and even made by flexible materials, making it a versatile technology.

However, plasma jets typically have a more focused treatment area but can penetrate deeper into tissue. Plasma jets also usually require higher gas flow rates, making it more challenging to regulate the flow field. Ultimately, the choice between DBD and plasma jets (or other cold plasma technologies) will depend on the specific application and desired treatment outcomes. Dielectric barrier discharge is a versatile and promising technology with a wide range of potential applications. Ongoing research is focused on developing new and innovative uses for DBD and improving its efficiency and effectiveness for existing applications.

References

[1] Racka-Szmidt K, Stonio B, Żelazko J, Filipiak M and Sochacki M 2022 A review: inductively coupled plasma reactive ion etching of silicon carbide *Materials* **15** 123

[2] Mozhayeva D and Engelhard C 2020 A critical review of single particle inductively coupled plasma mass spectrometry—A step towards an ideal method for nanomaterial characterization *J. Anal. At. Spectrom.* **35** 1740–83

[3] Keidar M and Beilis I 2013 *Plasma Engineering: Applications from Aerospace to Bio and Nanotechnology* (New York: Academic)

[4] Han X, Wei X, Xu H, Zhang W, Li Y, Li Y and Yang Z 2019 Investigation on the parameter distribution of Ar/O_2 inductively coupled plasmas *Vacuum* **168** 108821

[5] Lin L and Keidar M 2021 A map of control for cold atmospheric plasma jets: from physical mechanisms to optimizations *Appl. Phys. Rev.* **8** 011306

[6] Murphy A B *et al* 2017 The 2017 plasma roadmap: low temperature plasma science and technology *J. Phys. D: Appl. Phys.* **50** 323001

[7] Laroussi M *et al* 2022 Low-temperature plasma for biology, hygiene, and medicine: perspective and roadmap *IEEE Trans. Radiat. Plasma Med. Sci.* **6** 127–57

[8] Lu X P and Ostrikov K K 2018 Guided ionization waves: the physics of repeatability *Appl. Phys. Rev.* **5** 031102

[9] Shashurin A, Shneider M N, Dogariu A, Miles R B and Keidar M 2010 Temporary-resolved measurement of electron density in small atmospheric plasmas *Appl. Phys. Lett.* **96** 171502

[10] Shashurin A, Shneider M N, Dogariu A, Miles R B and Keidar M 2009 Temporal behavior of cold atmospheric plasma jet *Appl. Phys. Lett.* **94** 231504

[11] Mericam-Bourdet N, Laroussi M, Begum A and Karakas E 2009 Experimental investigations of plasma bullets *J. Phys. D: Appl. Phys.* **42** 055207

[12] Xian Y, Lu X, Cao Y, Yang P, Xiong Q, Jiang Z and Pan Y 2009 On plasma bullet behavior *IEEE Trans. Plasma Sci.* **37** 2068–73

[13] Naidis G V 2011 Modelling of plasma bullet propagation along a helium jet in ambient air *J. Phys. D: Appl. Phys.* **44** 215203

[14] Fridman , Alexander A, Lawrence A, Routledge F, Kennedy L A and Square P 2004 *Plasma Physics and Engineering* (Boca Raton, FL: CRC Press LLC)

[15] Laroussi M 2015 Low-temperature plasma jet for biomedical applications: a review *IEEE Trans. Plasma Sci.* **43** 703–12

[16] Begum A, Laroussi M and Pervez M R 2013 A brief study on the ignition of the non-thermal atmospheric pressure plasma jet from a double dielectric barrier configured plasma pencil *Plasma Sci. Technol* **15** 627–34

[17] Xiong Q, Nikiforov A Y, Lu X P and Leys C 2010 High-speed dispersed photographing of an open-air argon plasma plume by a grating-ICCD camera system *J. Phys. D: Appl. Phys.* **43** 415201

[18] Robert E, Darny T, Dozias S, Iseni S and Pouvesle J M 2015 New insights on the propagation of pulsed atmospheric plasma streams: from single jet to multi jet arrays *Phys. Plasmas* **22** 122007

[19] Kim J Y, Ballato J and Kim S O 2012 Intense and energetic atmospheric pressure plasma jet arrays *Plasma Processes Polym.* **9** 253–60

[20] Pei X, Wang Z, Huang Q, Wu S and Lu X 2011 Dynamics of a plasma jet array *IEEE Trans. Plasma Sci.* **39** 2276–7

[21] Li D, Liu D, Chen Z, Rong M and Kong M G 2016 A new plasma jet array source: discharge characteristics and mechanism *IEEE Trans. Plasma Sci.* **44** 2648–52

[22] Johnson M J, Brown G H, Boris D R, Petrova T B and Walton S G 2022 Two atmospheric pressure plasma jets driven by phase-shifted voltages: a method to control plasma properties at the plasma–surface interface *IEEE Trans. Plasma Sci.* **50** 2961–71

[23] Cho G, Kim J, Kang H, Kim Y, Kwon G C and Uhm H S 2012 Electrical potential measurement in plasma columns of atmospheric plasma jets *J. Appl. Phys.* **112** 103305

[24] Waskoenig J, Niemi K, Knake N, Graham L M, Reuter S, Schulz-Von Der Gathen V and Gans T 2010 Atomic oxygen formation in a radio-frequency driven micro-atmospheric pressure plasma jet *Plasma Sources Sci. Technol.* **19** 045018

[25] Hong Y C, Kang W S, Hong Y B, Yi W J and Uhm H S 2009 Atmospheric pressure air-plasma jet evolved from microdischarges: eradication of *E. coli* with the jet *Phys. Plasmas* **16** 123502

[26] Benedikt J, Focke K, Yanguas-Gil A and Von Keudell A 2006 Atmospheric pressure microplasma jet as a depositing tool *Appl. Phys. Lett.* **89** 251504

[27] Knake N, Reuter S, Niemi K, Schulz-Von Der Gathen V and Winter J 2008 Absolute atomic oxygen density distributions in the effluent of a microscale atmospheric pressure plasma jet *J. Phys. D: Appl. Phys.* **41** 194006

[28] Chen Z, Simonyan H, Cheng X, Gjika E, Lin L, Canady J, Sherman J H, Young C and Keidar M 2017 A novel micro cold atmospheric plasma device for glioblastoma both *in vitro* and *in vivo Cancers (Basel)* **9** 61

[29] Yao X, Yan D, Lin L, Sherman J H, Peters K B, Keir S T and Keidar M 2022 Cold plasma discharge tube enhances antitumoral efficacy of temozolomide *ACS Appl. Bio Mater.* **5** 1610–23
[30] Yao X, Lin L, Soni V, Gjika E, Sherman J H, Yan D and Keidar M 2020 Sensitization of glioblastoma cells to temozolomide by a helium gas discharge tube *Phys. Plasmas* **27** 114502
[31] Shi J J and Kong M G 2007 Radio-frequency dielectric-barrier glow discharges in atmospheric argon *Appl. Phys. Lett.* **90** 111502
[32] Walsh J L, Shi J J and Kong M G 2006 Submicrosecond pulsed atmospheric glow discharges sustained without dielectric barriers at kilohertz frequencies *Appl. Phys. Lett.* **89** 161505

Michael Keidar, Dayun Yan and Li Lin

Chapter 3

Diagnostics and modeling of cold atmospheric plasmas

Cold atmospheric plasmas (CAPs) are a partially ionized gas-like fluid that operates at near-ambient temperatures and pressures. They have attracted considerable attention in recent years due to their potential applications in various fields, such as materials processing, environmental remediation, biomedicine, and aerospace engineering. However, CAPs are also complex and dynamic systems that involve multiple physical and chemical processes, such as electron impact excitation and ionization, gas heating and cooling, plasma–surface interactions, and chemical reactions. Therefore, to understand and optimize CAPs for different purposes, it is essential to develop reliable and accurate diagnostic methods to provide information on plasma properties and behavior. Also, numerical simulation and modeling are important methods to study such a complex and dynamic system, while not all parameters are available through experimental diagnostics. The modeling can also be created through the simulation and verified by the experiments to help the researchers understand and discover the fundamental mechanisms behind the CAP behaviors.

3.1 General overview of CAP diagnostics

Plasma diagnostics is a crucial component of plasma research, providing information on the properties and behavior of plasma, including its temperature, density, and composition. In the context of CAP, used in various plasma medicine applications, plasma diagnostics play a critical role in understanding plasma chemistry and its interaction with biological materials. Plasma diagnostics can be classified into two main categories: invasive and non-invasive. Invasive diagnostics are active detection that involves inserting a probe or an electromagnetic beam into

doi:10.1088/978-0-7503-5537-7ch3

the plasma region, which may disturb the plasma equilibrium and introduce uncertainties in the measurements. Examples of invasive diagnostics are *Langmuir* probes and Faraday cups. There are also some non-invasive diagnostics that are also active detection, such as optical absorption spectroscopy (OAS), laser-induced fluorescence (LIF) [1–3], Thomson scattering (TS), and *Rayleigh* scattering (RS) [4, 5]. Passive non-invasive diagnostics rely on remote sensing techniques that do not interfere with the plasma state, such as optical emission spectroscopy (OES) and high-speed camera observations on streamer behaviors. Schlieren methods can either be passive such as the direct observation of gas flow using a visible light source, or active such as laser Schlieren deflectometry (LSD) [6–8].

One of the simplest diagnostic techniques is the *Langmuir* probe. *Langmuir* probes are widely used in plasma physics and diagnostics to measure the most important plasma parameters: the density and temperatures of electrons and ions. The general concept behind *Langmuir* probes is to use a small electrode to measure the current flowing between the electrode and the plasma, which is proportional to the density and energy of the plasma particles [1, 9, 10]. The electrode is typically cylindrical or spherical, with a diameter on the order of a few millimeters, and is biased with respect to the plasma potential. When the electrode is inserted into the plasma, it collects electrons and ions from the surrounding plasma, creating a current that flows through the electrode. There are many ways to use the *Langmuir* probe, such as a single probe and a multi-probe combination. During a measurement, the *Langmuir* probe provides a current signal when the user applies a DC voltage sweep. The density and temperatures of electrons and ions are thus computed based on the plasma sheath theory. The plasma sheath is a layer separating the bulk plasma from the electrode. The sheath is where the plasma potential drops from the bulk plasma potential to the electrode potential. The potential gradient, in other words, the electric field, accelerates the charged particles, creating the current measured by the *Langmuir* probe. However, this also leads to *Langmuir* probes' limitations in that the size of the sheath can be comparable to the CAP characteristic size, such as the diameter of a CAP jet. Therefore, the *Langmuir* probe disturbs the CAP too much to be an effective method for CAP diagnostics. Similarly, the Faraday cup is also not a proper method for CAP diagnostics. In addition, the interpretation of *Langmuir* probe measurements in collisional plasma is not trivial.

Considering these limits, one can conclude that effective immersive methods for CAP must rely on electromagnetic interactions. In the following sections, we will introduce the applications of LIF, TS, RS, and other non-immersive methods such as OES and Schlieren methods.

3.1.1 Laser-induced fluorescence

Laser-induced fluorescence (LIF) is a plasma diagnostic technique that is used to measure the concentration, distribution, and behavior of specific species in plasma, including reactive oxygen and nitrogen species (RONS) that are important in plasma

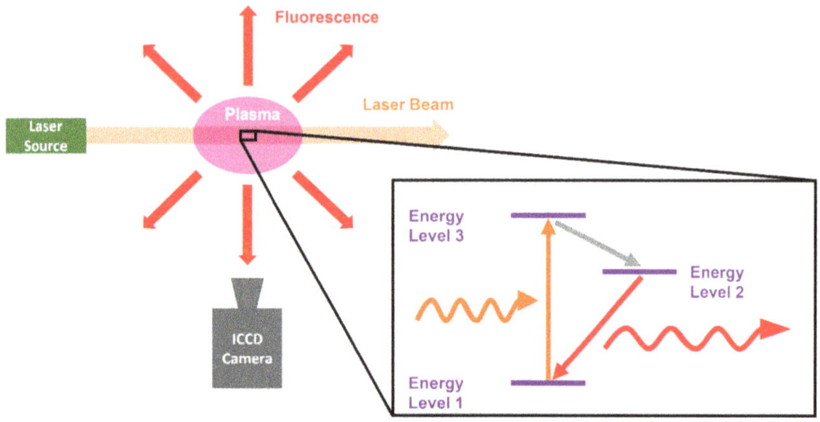

Figure 3.1. Applying laser-induced fluorescence on plasma.

medicine [11–14]. LIF excites the target species with a laser and measures the fluorescence emission.

The LIF technique involves several steps. As shown in figure 3.1, first, a laser is tuned to a specific wavelength corresponding to the target species' energy level. The laser beam is then directed onto the plasma, and the target species absorbs the laser energy, becoming excited that an electron is elevated from energy level 1 to energy level 3. The excited species decay to their ground state, such as quenching collisions. This leads to the electron drop from energy level 3 to energy level 2. Finally, emitting fluorescent light and returns to energy level 1. Therefore, the fluorescence photon has a longer wavelength than the incoming laser. The emitted light is collected by an optical system and directed to a detector, such as a photomultiplier tube (PMT), which measures the intensity of the fluorescence. The intensity of the fluorescence emission is proportional to the concentration of the target species in the plasma. Since different chemical species have different energy level structures, by varying the laser wavelength and scanning across different wavelengths, the LIF technique can provide information on the spatial distribution and behavior of the target species in the plasma [3].

One advantage of LIF over other plasma diagnostic techniques, such as OES, is its selectivity and sensitivity. LIF can selectively target specific species in plasma, providing highly resolved spatial and temporal information on their behavior. LIF can also be highly sensitive, allowing for species detection at low concentrations in plasma. However, there are also limitations to the LIF technique, including the need for specific laser wavelengths and the potential for interference from other species in the plasma. LIF is typically a point measurement technique and may not provide a complete picture of plasma behavior.

3.1.2 *Thomson* scattering and *Rayleigh* scattering

When the incident laser wavelength does not match the energy levels, one should have significant scatterings. For plasmas, the major parts are *Rayleigh* scattering

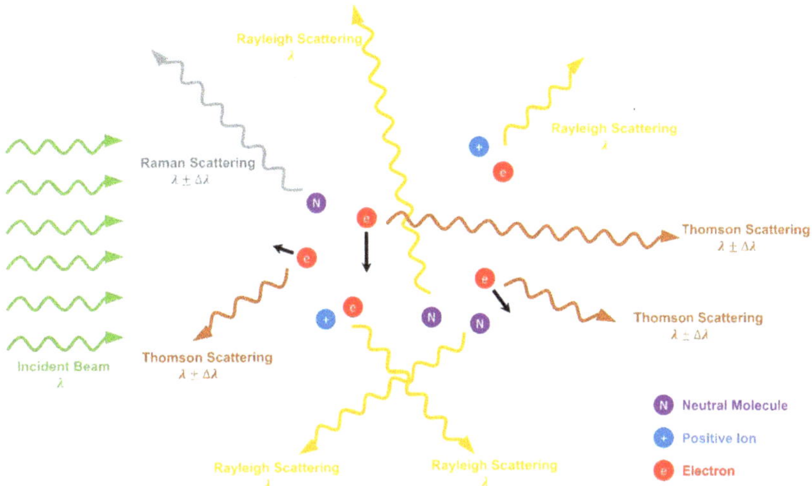

Figure 3.2. Electromagnetic beam scattering on a plasma. Three types of scattering might occur together in the plasma plume when the incident beam is a laser.

and *Thomson* scattering, along with a minor part of *Raman* scattering (figure 3.2). When an incident electromagnetic wave is applied to plasma, the electric field of the wave can oscillate the electrons. The bounded electrons, such as those in molecules and close to ions, have limited initial thermal velocity before the external wave impact. Therefore, they oscillate at the same frequency as the external wave. Such an oscillation is a forced plasma oscillation when the external frequency is higher than the plasma frequency. The oscillation then emits new electromagnetic waves at the same frequency in all directions. In a macroscopic view, the incident wave is scattered. When the target size is much smaller than the incident wavelength, such scattering is called *Rayleigh* scattering. *Rayleigh* scattering does not change the incident frequency; therefore, it is an elastic scattering. *Raman* scattering, however, involves the inelastic scattering of photons by molecules, which leads to a shift in the frequency of the scattered photons due to changes in the vibrational energy of the molecules. When an electromagnetic wave impacts plasmas, most of the photons result in *Rayleigh* scattering, with only very limited ones colliding on molecules can contribute to *Raman* scattering. On the other hand, the free electrons have significant thermal random thermal velocity (electron temperature), making a *Doppler* effect on the oscillation. The resulting scattering wavelength will contain a bandwidth surrounding incident one (figure 3.3).

Utilizing such a *Doppler* effect, laser *Thomson* scattering is a diagnostic technique used to measure the electron temperature of plasma (figure 3.4) [4]. The bandwidth represents the Doppler shift of the wavelength that can be measured and reversibly computed by the electron temperature as

$$T_e = A\frac{m_e c^2}{k_B}\left(\frac{\lambda_{\mathrm{FWHM}}}{\lambda_0}\right)^2 \tag{3.1}$$

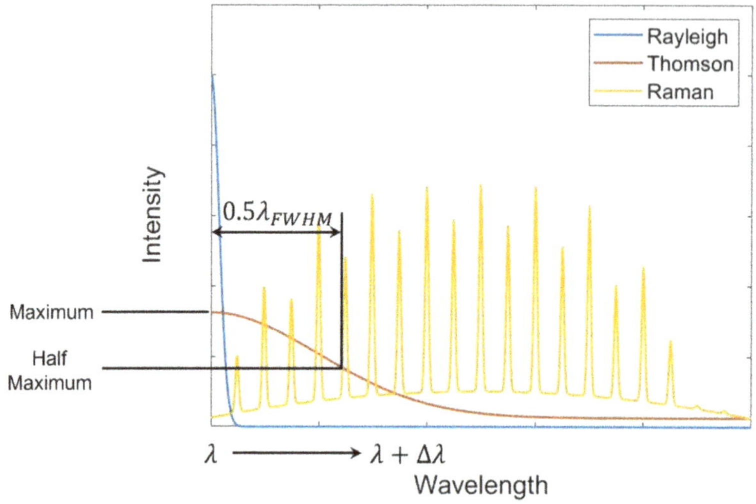

Figure 3.3. General concept of laser scattering signals of plasmas (not in scale). Please note that only the Stokes side is shown here for the Raman signal. The anti-Stokes side contains the peaks with a wavelength lower than λ.

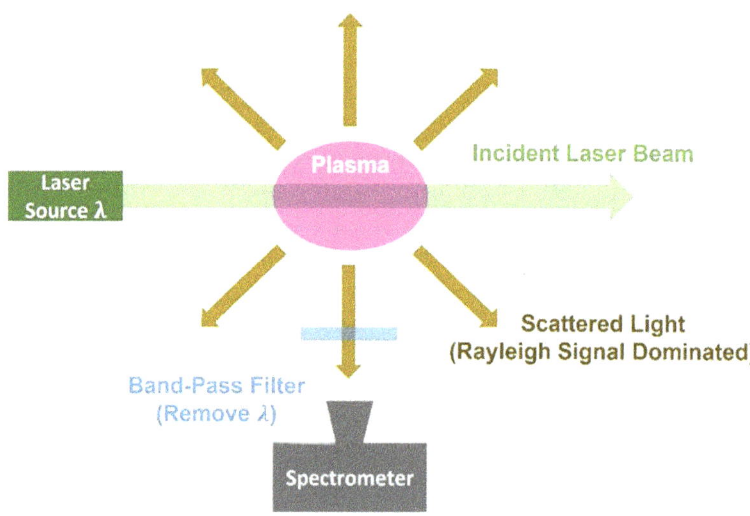

Figure 3.4. Laser *Thomson* scattering of plasma.

where T_e is the electron temperature, A is a coefficient determined by how the user makes a *Gaussian* fit on the *Thomson* signal, m_e is the mass of an electron, c is the light speed, k_B is the *Boltzmann* constant, λ_{FWHM} is the full width at half maximum of the *Thomson* peak (see figure 3.3), and λ_0 is the incident laser wavelength [15, 16]. One of the advantages of laser *Thomson* scattering is its ability to provide highly accurate measurements of electron temperature in a range of plasma conditions.

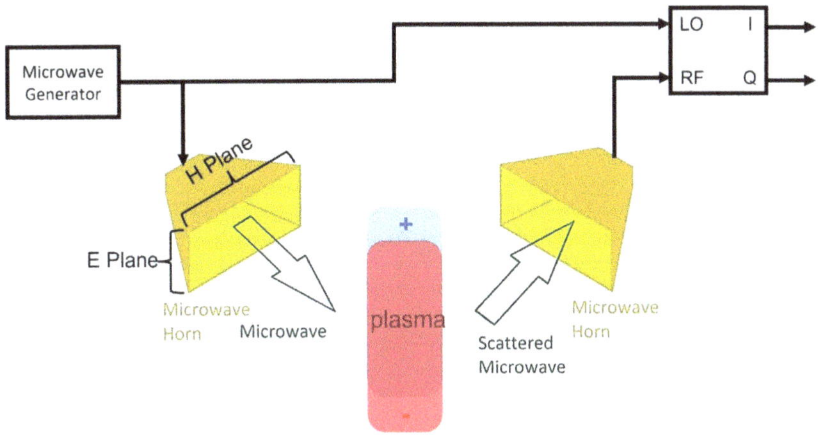

Figure 3.5. *Rayleigh* microwave scattering of plasma.

The method can also measure plasma density and velocity, providing a more complete picture of plasma dynamics. However, laser *Thomson* scattering also has limitations. The method can be technically challenging to implement, requiring careful calibration and alignment of the laser and optics. Additionally, the method is sensitive to environmental factors such as temperature and pressure, which can affect the accuracy of the measurements. Despite these challenges, laser *Thomson* scattering remains a valuable diagnostic tool for studying plasma dynamics and electron temperature in a wide range of applications, from laboratory experiments to fusion research and space plasma environments.

On the other hand, CAP usually contains an electron density at about 10^{13} cm^{-3}, this is a low ionization degree which leads to a low signal-to-noise ratio of laser Thomson scattering. Therefore, a higher-power laser is required to ensure better measurement. However, a higher laser power also may heat the plasma, and an excellent diagnostic method should not alter the sample too much. Recently, a *Rayleigh* microwave scattering method was developed to measure CAP's electron density and temperature. As shown in figure 3.5, a plasma oscillation is activated when an incoming microwave is hit on the CAP [5]. This bounded electron cloud oscillation emits *Rayleigh* scattering waves while its intensity is proportional to the electron density [5, 17]. However, the user needs to ensure that the plasma size is smaller than the incident microwave wavelength, as mentioned before, to agree with the definition of *Rayleigh* scattering definition. Otherwise, *Mie* scattering or geometric scattering will occur instead. Plasma diagnostics based on these two types of scattering is not yet developed. Therefore, the microwave used for *Rayleigh* scattering diagnostics requires a microwave wavelength long enough (frequency low enough) but still in a good beam shape. For example, a microwave above 10 GHz can be applied to a plasma plume up to about 3 cm. But since such a frequency is much higher than the sampling frequency of an oscilloscope or any other data acquisition instrument, one needs to modulate such a signal [18].

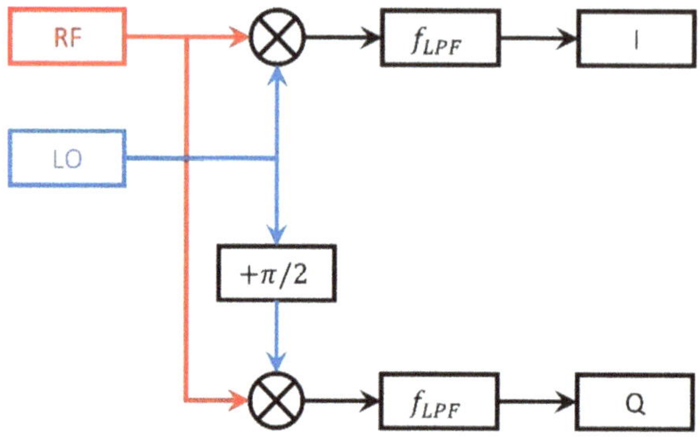

Figure 3.6. I/Q demodulation for the *Rayleigh* microwave signal.

The method shown in figure 3.5 is I/Q modulation and demodulation. First, let's consider a microwave generated from a microwave generator:

$$\mathrm{LO}(t) = E_{\mathrm{LO}}\cos(\omega_{\mathrm{LO}}t) \tag{3.2}$$

where LO stands for 'local oscillator' as the original microwave from the microwave generator, t is the time, E_{LO} is the electric field amplitude, ω_{LO} is the angular frequency. Then, the microwave will be emitted by a horn antenna toward the targeting plasma and scattered. The scattered wave, similarly, can be considered as:

$$\mathrm{RF}(t) = E_{\mathrm{RF}}(\mathrm{t})\cos(\omega_{\mathrm{RF}}t) \tag{3.3}$$

where RF stands for 'radio frequency' as the scattered wave with an amplitude at E_{RF} and frequency at ω_{RF}. The plasma oscillation will oscillate along with the E plane of the microwave. Once the scattered wave (RF) is collected by the receiver horn antenna, the LO and RF signal will be I/Q demodulated as shown in figure 3.6. The output will be the 'I' signal and 'Q' signal, which stand for 'in-phase' and 'quadrature'. First, the RF signal will be multiplied by the LO signal in the time domain. According to the trigonometric identity that $\cos\alpha\cos\beta = \frac{1}{2}[\cos(\alpha - \beta) + \cos(\alpha + \beta)]$, one will see that the results of RF multiplied by LO will result in a signal containing two frequencies. After the removal of the higher-frequency one by a low-pass-filter, one acquired the I signal as:

$$I(t) = f_{\mathrm{LPF}}[\mathrm{RF}(t)\mathrm{LO}(t)] = \frac{1}{2}E_{\mathrm{RF}}E_{\mathrm{LO}}\cos(\omega_{\mathrm{LO}}t - \omega_{\mathrm{RF}}t) \tag{3.4}$$

where $I(t)$ is the I signal, and f_{LPF} is the function of low-pass-filter. Similarly, we can have the Q signal by multiplying the RF with the LO after the LO signal is added a 90-degree phase shift, denoted as $\mathrm{LO}'(t)$:

$$\mathrm{LO}'(t) = E_{\mathrm{LO}}\sin(\omega_{\mathrm{LO}}t) \tag{3.5}$$

Then, the Q signal will be:

$$Q(t) = f_{\mathrm{LPF}}[\mathrm{RF}(t)\mathrm{LO}'(t)] = \frac{1}{2}E_{\mathrm{RF}}E_{\mathrm{LO}}\sin(\omega_{\mathrm{LO}}t - \omega_{\mathrm{RF}}t) \tag{3.6}$$

Therefore, one can see that the envelope of the RF signal can be acquired as:

$$U(t) = \sqrt{I^2 + Q^2} = \frac{\sqrt{2}}{2} E_{\mathrm{RF}}(t) E_{\mathrm{LO}} \tag{3.7}$$

where $U(t)$ is the envelope of RF signal $E_{\mathrm{RF}}(t)$multiplied by a known-constant $\frac{\sqrt{2}}{2} E_{\mathrm{LO}}$. This is not at a GHz frequency and, therefore can be measured by an oscilloscope. The amplitude E_{RF} is proportional to the electron density in the plasma and the horn positions. These parameters can be denoted as a calibration coefficient A and the math relation between U and the electron density n_e is:

$$U = A\sigma V \tag{3.8a}$$

$$\sigma = 2.82 \times 10^{-4} \frac{n_e \upsilon_m}{\omega^2 + \upsilon_m^2} \tag{3.8b}$$

where σ is the conductivity of the plasma, V is the plasma volume, ω is the incident microwave angular frequency and υ_m is the electron-neutral collision frequency [18, 19]. However, before using these equations, the calibration coefficient A needs to be quantified by using a material with known volume and dielectric constant based on the equation:

$$U = \varepsilon_0(\varepsilon_r - 1)\omega_{\mathrm{LF}} V \tag{3.9}$$

where ε_0 and ε_r is the dielectric constant of the vacuum and the material.

With the temporally resolved electron density known and the kinetic scheme of plasma chemistry known of CAP in the open air, one can also reversibly compute the electron temperature that can lead to such a temporally resolved electron density [20, 21]. This method will be discussed after the section of kinetic scheme at the end of chapter 3.

3.1.3 *Schlieren* methods

Schlieren methods are a group of diagnostic techniques used to visualize and quantify small variations in the refractive index of a medium, such as air, caused by variations in density, temperature, or composition. As shown in figure 3.7, the basic principle of a *Schlieren* system is to use a collimated light source to illuminate the medium under study and then to image the light passing through the medium onto a detector, such as a screen or a camera. A knife edge or other similar device is placed in front of the detector, blocking out the direct light, and only allowing the light that has been refracted by the medium to pass through. The distribution of different chemical species, gas density, and temperature can lead to a distribution of refractive index fields. Therefore, the refractive index variations in the medium can be visualized and quantified by measuring the intensity of the light passing through the knife edge. *Schlieren* methods are particularly useful for studying shock waves and other transient phenomena, as they can capture the rapid changes in the refractive index caused by the passage of a shock wave. In plasma physics, *Schlieren* methods have been used to study

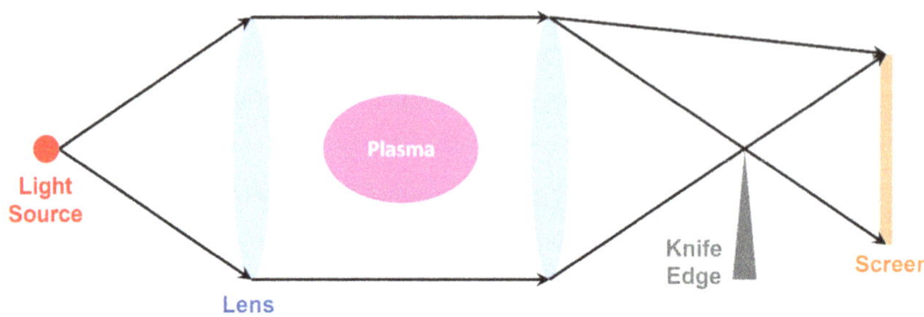

Figure 3.7. The hardware setup of a *Schlieren* measurement of plasma flow field using a regular light source.

shock waves in plasma jets and the behavior of plasma bubbles and plumes. Please note that *Schlieren* methods can also be combined with other diagnostic techniques, such as interferometry or laser-induced fluorescence, to provide a complete picture of plasma behavior. However, *Schlieren*'s methods also have limitations. They are typically only sensitive to variations in the refractive index along a single direction and may not provide a complete picture of the plasma behavior. Additionally, *Schlieren* methods can be sensitive to environmental factors such as temperature and humidity, which can affect the refractive index of the medium and introduce measurement errors (figure 3.7).

An example of *Schlieren*'s measurement of the plasma flow field can be found in [22]. It is clear that the plasma can lead to more turbulence in the flow field, which other publications also reported. This typical and famous observation of the CAP jet can only be discovered using the *Schlieren* method. More importantly, *Schlieren* can provide the downward CAP jet flow field because the CAP jet is usually used downward to treat liquid cell-medium targets *in vitro* or even living tissue during surgery. For example, the density of helium is about 10% of air at standard conditions. Therefore, a downward-pointing nozzle means that the helium can hardly reach too far down. However, when a plasma jet can be attracted and strengthened to a grounded target reported by other publications, the Schlieren observation also shows that the gas flow rate can also be compared more directly under such a condition [23, 24].

3.1.4 Optical emission spectrum (OES)

OES is a commonly used plasma diagnostic technique that measures the spectrum of light emitted by the plasma [25–29]. The OES technique is based on the fact that atoms and molecules in plasma emit light at specific wavelengths, corresponding to the emitting species' energy levels. OES can indirectly provide qualitative and quantitative information on plasma composition, excitation temperature, electron density, and plasma dynamics. However, OES also has some limitations and challenges, such as spectral interference, self-absorption, calibration, and, most importantly, the interpretation of the emission spectra. Various methods have been developed and applied to overcome these difficulties, such as spectral fitting, line-ratio method, collisional-radiative model, Abel inversion, and machine learning

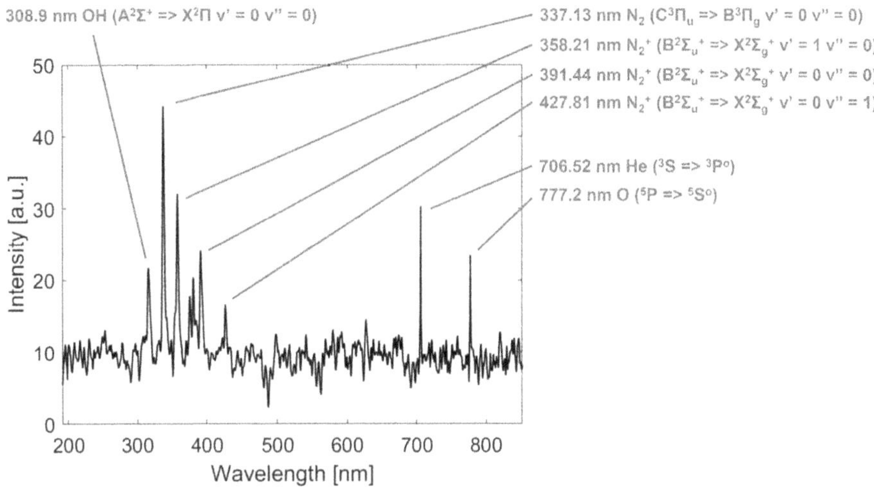

Figure 3.8. An example of an optical emission spectrum of a plasma jet containing helium and air species.

techniques. Overall, OES is a versatile and powerful tool for CAP diagnostics, especially when combined with other optical or electrical methods. The detailed physical mechanism behind the emission of light by a CAP is spontaneous emissions. In general, a CAP emits light due to the excitation of electrons and ions in the plasma. When an electron or ion collides with an atom or molecule in the plasma, it can transfer energy to the atom or molecule, causing it to become excited or ionized. The excited species can then emit photons with specific wavelengths as they return to their ground state or lower energy levels. The emitted light thus contains a combination of many wavelengths, usually in the UV–vis range. The spectrum is then collected by an optical system, typically a fiber optic cable, and directed to a spectrometer for analysis. The spectrometer separates the emitted light into its constituent wavelengths and measures the intensity of the light at each wavelength. The resulting spectrum is a plot of intensity versus wavelength and provides information on the composition and temperature of the plasma.

A typical OES result of a helium–air CAP is shown in figure 3.8. This explains the purple color of the CAP jet, and most of the highest peaks are from the $N_2(C)$ => $N_2(B)$, which is known as the second positive system, containing several peaks with different vibrational quantum numbers [30]. There are other famous air species peaks, such as the 308.9 nm OH and the 777.2 nm atomic O. OES is usually used to verify species' existence but can hardly quantify the concentrations. When a single species dominates, one can also use a certain peak ratio to compute the electron temperature [31].

3.2 Numerical simulation and modeling

3.2.1 General concept of finite different simulation and particle-in-cell algorithm

Low-temperature atmospheric plasma physics and chemistry is a rapidly growing field that has gained attention over the past few decades due to its numerous applications in science, engineering, and medicine. Plasma, a gas composed of charged particles, has long been studied at high temperatures in fields such as fusion research, but in recent years, low-temperature plasma has emerged as a new area of interest due to its unique properties and potential applications. Numerical simulations and modeling play an essential role in understanding and predicting the behavior of low-temperature plasma, which can be challenging to observe directly due to its complex nature. In recent years, there has been a growing interest in low-temperature plasma, which operates at atmospheric pressure and has potential applications in plasma medicine, surface modification, and environmental remediation. Numerical simulations and modeling have played a critical role in advancing research in low-temperature plasma by providing insights into plasma dynamics, chemistry, and interactions between plasma and materials.

However, CAP is a plasma exposed in the open air; this means the complexity of plasma chemistry contains hundreds of chemical species with thousands of chemical reactions co-occurring, each having a unique and dynamic reaction rate coefficient, usually a function of reactant temperature. This means that the CAP simulation should contain a complex kinetic scheme. The chemical reactions, in other words, the inelastic collisions among particles, are usually simulated using the *Monte Carlo* method, as the right-hand side of the mass equation, which will be discussed below. The *Monte Carlo* method is a statistical method used to simulate the behavior of individual particles in plasma [32]. This method simulates the probability of particle interactions, such as electron scattering and ionization, and can be used to investigate the plasma chemistry and ionization processes in CAP.

Like all other simulations, the simulation of CAP is a loop of the energy equation, momentum equation, and mass equation:

$$\nabla \boldsymbol{E} = -\nabla^2 \varphi = \frac{q}{\varepsilon} \tag{3.10}$$

$$T_e = f\left(\frac{|\boldsymbol{E}|}{n_e}\right) \tag{3.11}$$

$$\frac{\partial \boldsymbol{v}_e}{\partial t} = -e\boldsymbol{E} + \boldsymbol{F}_c \tag{3.12}$$

$$\frac{\partial n_e}{\partial t} + \nabla \cdot (n_e \boldsymbol{v}_e) = \frac{\partial G_e}{\partial t} \tag{3.13}$$

$$\frac{\partial G_\zeta}{\partial t} = \sum_\eta k_\eta \prod_\xi n_{\eta,\xi}^{\Delta} \tag{3.14}$$

where $\boldsymbol{E}$ is the electric field, φ is the electric potential, q is the net charge density, ε is the permittivity, T_e is the electron temperature, $\boldsymbol{F}_c$ is the drag due to the collisions, n_e is the electron density, t is time, $\boldsymbol{v}_e$ is the drift velocity of the electron, e is the unit charge, G_e is the gain of electron density, G_ζ is the gain of the ζth chemical species, k_η

is the reaction rate coefficient of the ηth reaction generating or consuming the ζth species, and $n_{\eta,\xi}^{\Delta}$ is the reactant density of the ξth reactant in the ηth reaction with a reaction order Δ. Equation (3.10) is the Poisson equation that can solve the electric field from the boundary condition and the distribution of charges. Because of the weak current in CAP, unless the users apply an external magnetic field, the model of CAP usually does not include a magnetic field; thus, the other three *Maxwell* equations are not required. The reduced electric field $\frac{|E|}{n_e}$ can therefore be used to solve for the mean electron temperature T_e and the electron energy distribution function (EEDF) depending on the model used to describe the function $f\left(\frac{|E|}{n_e}\right)$. Equations (3.10) and (3.11) are thus together serving as an energy equation in the model. Equation (3.12) is the momentum equation to solve for the electron drift velocity. The magnetic field is missing, but the drag of collisions can be contained. Equation (3.13) is the electron's mass equation, a.k.a. continuity equation. The right-hand side is the net gain or loss of the electron density due to the chemical reactions that are described by the general form of rate equation (3.14). Please note that equation (3.13) maybe not be directly solved but can also be achieved by using the famous particle-in-cell algorithm (PIC) [9, 10]. As shown in figure 3.9, each electron density in a meshed cell is considered as a single super electron that can be moved by the drift velocity v_e. Once the super electron is moved to a new location, its density will be divided into four parts and placed into the four neighboring cells: A, B, C, and D as shown in the figure. The dividing ratio can be either determined by the area ratio or the distance (from the super electron location to the four cell centers) ratio. This process is also known as 'weighting'. The weighting process can be performed using various interpolation schemes, such as linear or quadratic interpolation, depending on the desired level of accuracy and computational efficiency. In general, higher-order interpolation schemes provide greater accuracy but require more computational resources. The rate equation (3.14) can either be solved in matrix operations or using a *Monte Carlo* algorithm that uses random numbers in view of statistical meaning to determine if a chemical reaction or a collision can occur. By combining the *Monte Carlo* method with the PIC algorithm, it is possible to model the effects of collisions on the behavior of the plasma in a computationally efficient manner. This approach has been used in a wide range of applications, including fusion research, plasma processing, and plasma medicine (figure 3.9).

Figure 3.10 shows a snapshot of a PIC simulation of a helium-guided CAP jet as an example. The subplots are all contour plots of the time at 42 ns after the generation of the streamer. The geometry setup is a typical pin–ring design of a CAP jet which can be found in chapter 2. The streamer head can be found in the plot of ionization rate where a yellow area is located at Z = 1.5 cm. This is also the wavefront of an ionization wave which is the region of an ionization rate peak. Therefore, The electron density continuously decreases from the anode (Z = 0.25 cm and R = 0 cm) to the streamer. This is also true for the electric potential, but the potential only decreases significantly in the vicinity of the streamer head. This

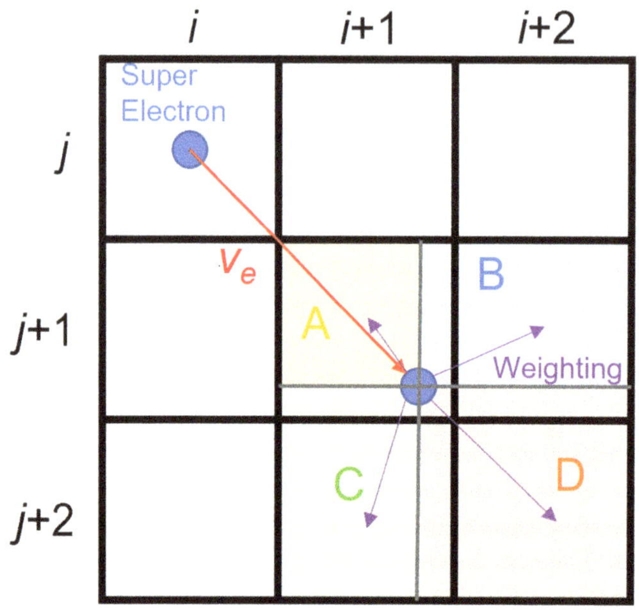

Figure 3.9. The key concept of PIC algorithm: the weighting of a super particle.

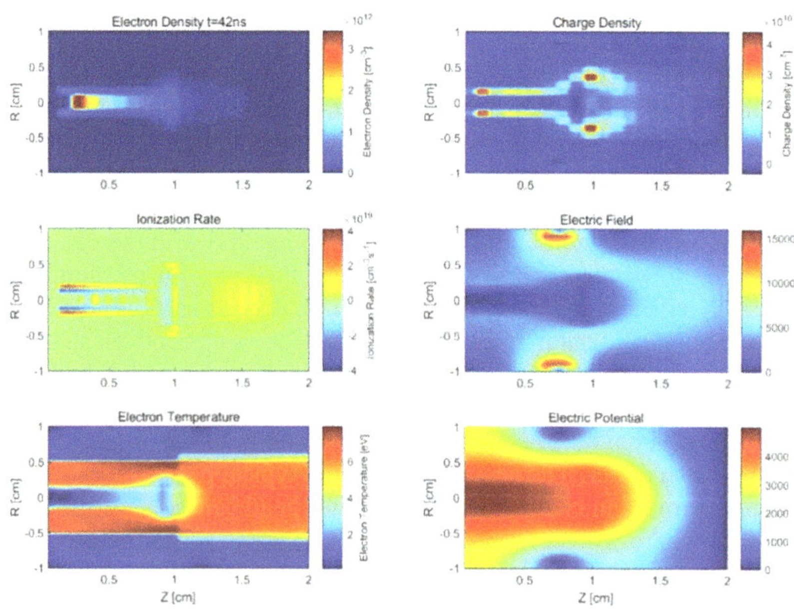

Figure 3.10. A snapshot of a 2D PIC simulation of a helium plasma jet in ambient air.

feature reflects the common sense that the plasma column has a high conductivity where the electric field is low, as shown in the contour plot of the electric field. The electron temperature is proportional to the electric field. In the net charge plot, one can also find that the streamer head is positively charged with a negatively charged seed electron on the right and a quasi-neutral plasma column on the left. All these features agree with what we introduced in chapter 2 and other plasma jet simulation papers [33, 34].

Overall, experimental measurements of cold atmospheric plasma have limitations in acquiring all the desired parameters, and numerical simulations have become an essential tool for understanding the plasma and obtaining the unavailable data. In addition to verifying physical models, numerical simulations can also train machine learning systems that can be used to analyze plasma and optimize the control of plasma generators. This allows for the development of automatic control and optimization of plasma-based processes, which can significantly enhance their efficiency and effectiveness. As such, numerical simulations are critical for advancing our understanding and utilization of cold atmospheric plasma in various fields, including plasma medicine, material processing, and environmental remediation.

3.2.2 Kinetic scheme and modeling-assisted diagnostics

The kinetic scheme of cold atmospheric plasma is complicated. It includes at least a noble gas and air species. The combination of N, O, and H elements from the humidity can lead to hundreds of species with different excited states and photons with different wavelengths. The chemical reactions involve ionization, dissociation, attachment, association, and other photon-related processes [35, 36].

Ionization is the most fundamental reaction to generate plasma. It is the process of removing or adding one or more electrons to an atom or molecule, forming ions. In CAPs, ionization can occur through electron impact, where an electron collides with a gas molecule and removes one or more electrons from it, or through photoionization, where a photon with enough energy ionizes a gas molecule. Ionization is crucial in sustaining plasma by providing the electrons needed for further reactions. The ionization threshold is the minimum energy required to ionize an atom or molecule. In other words, it is the energy needed to remove an electron from the atom or molecule, creating a positively charged ion. This energy can be supplied by various means, such as collisions with other particles or electromagnetic radiation. The energy required for ionization varies depending on the species being ionized and the ionization method. For example, helium's ionization threshold is higher than hydrogen's because helium has a larger atomic number and more tightly bound electrons. The ionization rate is the rate at which neutral particles in plasma are converted into ions by various ionization processes. The ionization rate can be computed by the rate equation (3.17). The rate coefficient is a function of the species temperature, especially the electron temperature for the electron impact ionization. The rate coefficient is boosted when the electron temperature exceeds the threshold (figure 3.11) [37]. Other types of chemical reactions can have similar features, known

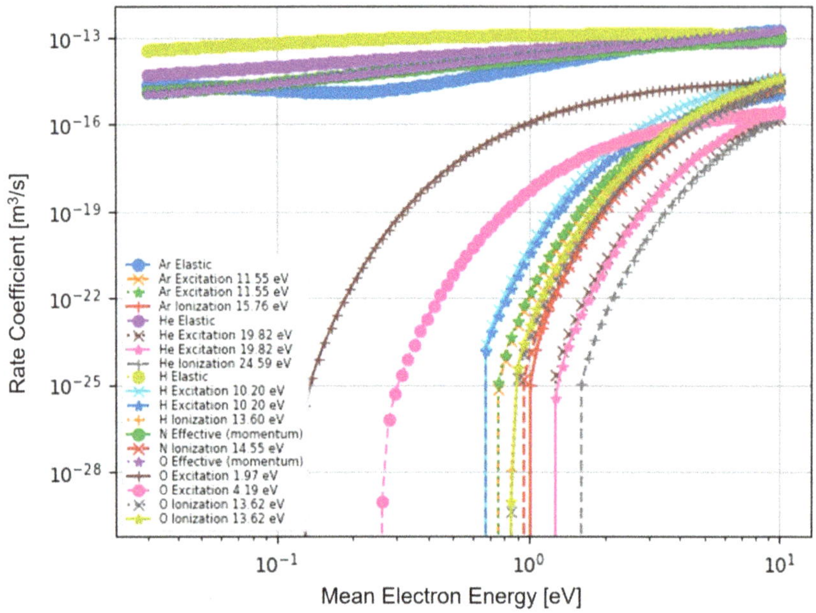

Figure 3.11. Rate coefficients of elastic collisions and inelastic collisions of electron impacts on Ar, He, H, N, and O as functions of the mean electron energy.

as the *Arrhenius* equation. In general, the ionization rate is proportional to the plasma density and the rate at which ionizing collisions occur.

Dissociation is the process of breaking a molecule into two or more fragments. In CAPs, dissociation can occur through electron impact or photon absorption. For example, N_2 can dissociate into N and N by absorbing an electron or a photon. Dissociation is essential for producing reactive species, such as radicals, which can participate in further reactions.

Attachment is the process of capturing an electron by a neutral molecule or atom, forming a negative ion. In CAPs, attachment can occur through electron impact, where an electron attaches to a neutral molecule or atom, or associative detachment, where a neutral molecule or atom collides with a negative ion and captures the extra electron. Attachment is an important process for producing negative ions, which can also participate in further reactions.

The association is the process of combining two or more neutral species to form a new molecule or compound. In CAPs, the association can occur through reactions such as N + O => NO or H + O_2 => HO_2. The association is essential for producing stable molecules and compounds, which can then be released into the surrounding environment.

Penning ionization is a process in which a metastable species transfers its energy to a neutral molecule, resulting in the ionization of the neutral molecule. Penning ionization can occur in CAPs when a metastable species such as He* collides with a neutral molecule such as O_2, resulting in the ionization of O_2. *Penning* ionization is vital for producing ions in CAPs [38, 39].

Massive publications contain the kinetic scheme of the reactive species generated in CAP, and their pathways [6]. Based on fully understanding the CAP's chemical model, one can combine it with other diagnostics to measure electron temperature using *Rayleigh* microwave scattering [20]. Considering the continuity equation of the electron density

$$\frac{\partial N_e}{\partial t} + \int \nabla \cdot \frac{\boldsymbol{j}_e}{e} \mathrm{d}v = R_{\text{gain}} - R_{\text{loss}} \tag{3.15a}$$

$$\frac{\boldsymbol{j}_e}{e} = D\nabla n_e + \mu \boldsymbol{E} n_e = \frac{k_B T_e}{m_e v_{\text{coll}}} \nabla n_e + \frac{eE}{m_e v_{\text{coll}}} n_e \tag{3.15b}$$

where N_e is the total electron number, $\boldsymbol{j}_e$ is the electron current, e is the unit charge, $\mathrm{d}v$ is the unit volume, the right-hand-side $R_{\text{gain}} - R_{\text{loss}}$ is the net gain of the electron number determined by the kinetic scheme, D is the diffusion coefficient, μ is the mobility, m_e is the mass of a single electron, and v_{coll} is the electron collision frequency. In the continuity equation, the electron density n_e can be measured by the *Rayleigh* microwave scattering, the electric field $\boldsymbol{E}$ can be measured by the streamer potential [40–42], and the size of plasma filament [43–45]. All other parameters are constants, and the only unknown, the electron temperature T_e can therefore be solved.

References

[1] Merlino R L 2007 Understanding Langmuir probe current–voltage characteristics *Am. J. Phys.* **75** 1078–85

[2] Schmidt-Bleker A, Winter J, Iseni S, Dünnbier M, Weltmann K D and Reuter S 2014 Reactive species output of a plasma jet with a shielding gas device—combination of FTIR absorption spectroscopy and gas phase modelling *J. Phys. D: Appl. Phys.* **47** 145201

[3] Riès D, Dilecce G, Robert E, Ambrico P F, Dozias S and Pouvesle J M 2014 LIF and fast imaging plasma jet characterization relevant for NTP biomedical applications *J. Phys. D: Appl. Phys.* **47** 275401

[4] Muraoka K and Kono A 2011 Laser thomson scattering for low-temperature plasmas *J. Phys. D: Appl. Phys.* **44** 043001

[5] Shneider M N and Miles R B 2005 Microwave diagnostics of small plasma objects *J. Appl. Phys.* **98** 033301

[6] Lin L and Keidar M 2021 A map of control for cold atmospheric plasma jets: from physical mechanisms to optimizations *Appl. Phys. Rev.* **8** 011306

[7] Lu X, Naidis G V, Laroussi M and Ostrikov K 2014 Guided ionization waves: theory and experiments *Phys. Rep.* **540** 123–66

[8] Lu X P and Ostrikov K K 2018 Guided ionization waves: the physics of repeatability *Appl. Phys. Rev.* **5** 031102

[9] Keidar M and Beilis I 2013 *Plasma Engineering: Applications from Aerospace to Bio and Nanotechnology* (New York: Academic)

[10] Fridman , Alexander A, Lawrence A, Routledge F, Kennedy L A and Square P 2004 *Plasma Physics and Engineering* (Boca Raton, FL: CRC Press)

[11] Pei X, Wu S, Xian Y, Lu X and Pan Y 2014 On OH density of an atmospheric pressure plasma jet by laser-induced fluorescence *IEEE Trans. Plasma Sci.* **42** 1206–10
[12] Amorim J, Baravian G and Jolly J 2000 Laser-induced resonance fluorescence as a diagnostic technique in non-thermal equilibrium plasmas *J. Phys. D: Appl. Phys.* **33** R51
[13] van den Bekerom D, Richards C, Huang E, Adamovich I and Frank J H 2022 2D imaging of absolute methyl concentrations in nanosecond pulsed plasma by photo-fragmentation laser-induced fluorescence *Plasma Sources Sci. Technol.* **31** 095018
[14] Wagenaars E, Gans T, O'Connell D and Niemi K 2012 Two-photon absorption laser-induced fluorescence measurements of atomic nitrogen in a radio-frequency atmospheric-pressure plasma jet *Plasma Sources Sci. Technol.* **21** 042002
[15] Van Der Meiden H J *et al* 2008 High sensitivity imaging Thomson scattering for low temperature plasma *Rev. Sci. Instrum.* **79** 013505
[16] Belostotskiy S G, Khandelwal R, Wang Q, Donnelly V M, Economou D J and Sadeghi N 2008 Measurement of electron temperature and density in an argon microdischarge by laser Thomson scattering *Appl. Phys. Lett.* **92** 221507
[17] Raizer Y P 1997 *Gas Discharge Physics* (Berlin: Springer)
[18] Shashurin A, Shneider M N, Dogariu A, Miles R B and Keidar M 2010 Temporary-resolved measurement of electron density in small atmospheric plasmas *Appl. Phys. Lett.* **96** 171502
[19] Lin L and Keidar M 2016 Cold atmospheric plasma jet in an axial DC electric field *Phys. Plasmas* **23** 083529
[20] Lin L, Lyu Y, Shneider M N and Keidar M 2018 Average electron temperature estimation of streamer discharge in ambient air *Rev. Sci. Instrum.* **89** 113502
[21] Lin L, Lyu Y, Trink B, Canady J and Keidar M 2019 Cold atmospheric helium plasma jet in humid air environment *J. Appl. Phys.* **125** 153301
[22] Robert E, Sarron V, Darny T, Riès D, Dozias S, Fontane J, Joly L and Pouvesle J M 2014 Rare gas flow structuration in plasma jet experiments *Plasma Sources Sci. Technol.* **23** 012003
[23] Schmidt-Bleker A, Reuter S and Weltmann K D 2015 Quantitative schlieren diagnostics for the determination of ambient species density, gas temperature and calorimetric power of cold atmospheric plasma jets *J. Phys. D: Appl. Phys.* **48** 175202
[24] Qaisrani M H, Xian Y, Li C, Pei X, Ghasemi M and Lu X 2016 Study on dynamics of the influence exerted by plasma on gas flow field in non-thermal atmospheric pressure plasma jet *Phys. Plasmas* **23** 063523
[25] Zhu X M, Walsh J L, Chen W C and Pu Y K 2012 Measurement of the temporal evolution of electron density in a nanosecond pulsed argon microplasma: using both Stark broadening and an OES line-ratio method *J. Phys. D: Appl. Phys.* **45** 295201
[26] Crintea D L, Czarnetzki U, Iordanova S, Koleva I and Luggenhölscher D 2009 Plasma diagnostics by optical emission spectroscopy on argon and comparison with Thomson scattering *J. Phys. D: Appl. Phys.* **42** 045208
[27] Puač N, Maletić D, Lazović S, Malović G, Dordević A and Petrović Z L J 2012 Time resolved optical emission images of an atmospheric pressure plasma jet with transparent electrodes *Appl. Phys. Lett.* **101** 024103
[28] Bruggeman P, Walsh J L, Schram D C, Leys C and Kong M G 2009 Time dependent optical emission spectroscopy of sub-microsecond pulsed plasmas in air with water cathode *Plasma Sources Sci. Technol.* **18** 045023

[29] Uchida G, Takenaka K, Kawabata K and Setsuhara Y 2015 Influence of He gas flow rate on optical emission characteristics in atmospheric dielectric-barrier-discharge plasma jet *IEEE Trans. Plasma Sci.* **43** 737–44

[30] Gaydon A G and Pearse R 1976 *The Identification of Molecular Spectra* (London: Chapman and Hall)

[31] Zhu X M and Pu Y K 2008 Using OES to determine electron temperature and density in low-pressure nitrogen and argon plasmas *Plasma Sources Sci. Technol.* **17** 024002

[32] Sommerer T J and Kushner M J 1992 Numerical investigation of the kinetics and chemistry of rf glow discharge plasmas sustained in He, N_2, O_2, He/N_2/O_2, He/CF_4/O_2, and SiH_4/NH_3 using a Monte Carlo-fluid hybrid model *J. Appl. Phys.* **71** 1654–73

[33] Brok W J M, Bowden M D, Van Dijk J, Van Der Mullen J J A M and Kroesen G M W 2005 Numerical description of discharge characteristics of the plasma needle *J. Appl. Phys.* **98** 013302

[34] Naidis G V 2017 Modeling of streamer dynamics in atmospheric-pressure air plasma jets *Plasma Processes Polym.* **14** 1600127

[35] Waskoenig J, Niemi K, Knake N, Graham L M, Reuter S, Schulz-Von Der Gathen V and Gans T 2010 Atomic oxygen formation in a radio-frequency driven micro-atmospheric pressure plasma jet *Plasma Sources Sci. Technol.* **19** 045018

[36] Norberg S A, Johnsen E and Kushner M J 2015 Formation of reactive oxygen and nitrogen species by repetitive negatively pulsed helium atmospheric pressure plasma jets propagating into humid air *Plasma Sources Sci. Technol.* **24** 35026

[37] Hagelaar G J M and Pitchford L C 2005 Solving the Boltzmann equation to obtain electron transport coefficients and rate coefficients for fluid models *Plasma Sources Sci. Technol.* **14** 722–33

[38] Miller W H 1970 Theory of penning lonization. I. Atoms *J. Chem. Phys.* **52** 3563–72

[39] Bell K L, Dalgarnot A and Kingston A E 1968 Penning ionization by metastable helium atoms *J. Phys. B: At. Mol. Phys.* **1** 18

[40] Shashurin A, Scott D, Shneider M N and Keidar M 2014 *IEEE Trans. Plasma Sci.* **42** 2402–3

[41] Cho G, Kim J, Kang H, Kim Y, Kwon G C and Uhm H S 2012 Electrical potential measurement in plasma columns of atmospheric plasma jets *J. Appl. Phys.* **112** 103305

[42] Shashurin A, Shneider M N and Keidar M 2012 Measurements of streamer head potential and conductivity of streamer column in cold nonequilibrium atmospheric plasmas *Plasma Sources Sci. Technol.* **21** 034006

[43] Lu X P *et al* 2009 A cold plasma cross made of three bullet-like plasma plumes *Thin Solid Films* **518** 967–70

[44] Mericam-Bourdet N, Laroussi M, Begum A and Karakas E 2009 Experimental investigations of plasma bullets *J. Phys. D: Appl. Phys.* **42** 055207

[45] Xian Y, Lu X, Cao Y, Yang P, Xiong Q, Jiang Z and Pan Y 2009 On plasma bullet behavior *IEEE Trans. Plasma Sci.* **37** 2068–73

Michael Keidar, Dayun Yan and Li Lin

Chapter 4

Chemically-based cold atmospheric plasma treatment

Investigating the different responses to cold atmospheric plasma (CAP) treatment is the key to understanding its anti-cancer mechanism. CAP, as a tunable source of reactive species that robustly exerts biochemical effects on mammalian cells, particularly on cancer cells. In chapter 4, we introduce the current state of the chemically based treatment *in vitro*. We focus on the general conclusions that have been drawn about cellular responses. Lastly, we specifically introduce some recent discoveries in plasma medicine which provide a unique perspective to understand the uniqueness of plasma medicine.

4.1 Anti-cancer demonstration *in vitro*

As early as 2004, researchers reported the first observation of apoptosis in CAP-treated lung cancer cells [1]. However, the authors did not systematically investigate this phenomenon. The first complete study about the cytotoxicity of DBD treatment on melanoma cells was done in 2007 [2]. Since then, CAP has shown strong cytotoxicity to more than 20 cancer types *in vitro*, including skin cancer, head & neck cancer, brain cancer, breast cancer, colorectal cancer, lung cancer, cervical cancer, hepatoma, as well as leukemia [3–11]. Evidence also shows that CAP has a strong capacity to restore the sensitivity of many chemo-resistant cancer cell lines, including tumor necrosis factor-related apoptosis-inducing ligand (TRAIL)-resistant colorectal cancer cells, temozolomide (TMZ)-resistant glioblastoma cells, paclitaxel-resistant breast cancer cells, and the 5-fluorouracil-resistant hepatocarcinoma cells [12–16]. More attractively, CAP's cytotoxicity to cancer cells shows high levels of selectivity in many studies [10, 17–21]. Chapter 7 will systematically discuss the selective anti-cancer properties of CAP treatment.

doi:10.1088/978-0-7503-5537-7ch4

Besides, CAP can effectively inhibit and kill cancer stem cells (CSCs). It is known that intratumoral heterogeneity presents a challenge to existing paradigms of anticancer therapy. The limitations of conventional anticancer therapies leading to treatment failure and cancer recurrence are mainly due to the drug resistance and the self-renewal capacities of CSCs. Addressing this, Trink *et al* studied the anti-cancer effect of CAP on prostate CSCs and demonstrated that two types of heterogeneous CSC populations derived from a single aggressive Gleason grade 9 tumor-bearing patient are sensitive to the effects of plasma treatment [22]. Surprisingly, the more aggressive CSC population (C13) was more sensitive to CAP treatment than the less aggressive type (C12). Experimental evidence suggests that CAP can be delivered in cytotoxic doses to CSCs [23]. These studies demonstrate the potential of CAP treatment to induce highly significant cytotoxic effects in cells that are thought to be a causal factor in treatment resistance and relapse.

CAP can be used alongside other cutting-edge cancer treatment modalities. The synergistic use of CAP with other existing approaches such as nanotherapy did improve the treatment's efficacy [24–31]. The cytotoxicity of CAP treatment on melanoma was enhanced by pretreating melanoma cancer cells with the anti-FAK antibody-conjugated gold (Au) nanoparticles [27]. The surface-modified nanoparticles weaken the function of FAK which may intensify the CAP-triggered detachment of melanoma cells from the substrate. We further demonstrated that the pretreatment of Au nanoparticles even without any specific modification could enhance the cytotoxicity of CAP jet treatment to glioblastoma cells (U87MG) [24]. CAP treatment changes the uptake of nanoparticles and molecules, which can explain the synergistic application of CAP and AuNPs [32]. Another study further demonstrated that the combined treatment of CAP and polyethylene glycol (PEG)-coated Au nanoparticles increased cancer cell death in solid tumors and decreased epithelial-mesenchymal transition (EMT) [33]. The drug encapsulated core-shell nanoparticles have also shown its synergistic anti-cancer capacity with CAP treatment on breast cancer cell lines [34].

4.2 Two treatment approaches

To date, studies have mainly been performed by treating cancer cell lines and corresponding homologous normal counterpart cell lines cultured in dishes, or multi-well plates, or flasks. Two general strategies for using CAP have been developed. One is to employ the bulk plasma from CAP source to directly treat the cells grown *in vitro* or the subcutaneous tumors grown in animal models, which exposes samples to CAP's chemical and physical factors simultaneously (figure 4.1(a)) [2, 35]. Direct treatment is the most conventional approach [2]. Alternatively, solutions such as cell culture medium, phosphate buffered saline (PBS), and others can be used as a carrier of long-lived reactive species to indirectly exert killing effects on cells or be injected into animals to affect tumorous tissues (figure 4.1(b)) [36, 37]. CAP-activated solutions (PAS) and their application will be

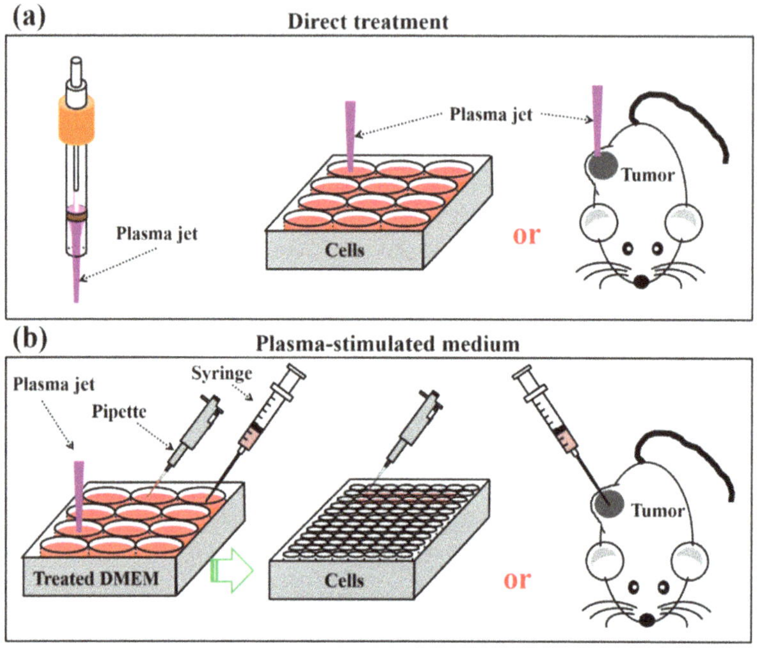

Figure 4.1. Research strategy of CAP cancer treatment. (a). Direct treatment on cancer cells *in vitro* or on subcutaneous xenografted tumors *in vivo*. (b). Indirect treatment on solution. CAP-activated solutions will be used to inhibit the growth of cancer cells *in vitro* and the tumor tissues in animals. Reproduced from [38], CC BY 3.0.

systematically introduced in chapter 5. Our discussion will focus on the direct CAP treatment in chapter 4.

Understanding the interaction between CAP and its target cells will build the foundation for using CAP in cancer therapy. Strong evidence supports that this interaction results from a combination of both chemical and physical factors. Chemical factors include dozens of reactive species that are generated in the gas phase of CAP. These reactive species may include long-lived components such as ions and short-lived components such as metastable radicals. In plasma medicine, scholars tend to classify long-lived and short-lived reactive species in terms of their chemical/physical properties.

Both oxygen-based species, such as hydroxyl ($OH^{\bullet}$), singlet oxygen ($^1O_2$), superoxide (O_2^-), hydrogen peroxide (H_2O_2), ozone (O_3), as well as nitrogen-based species, such as nitric oxide (NO), nitrogen dioxide (NO_2), nitrogen trioxide (NO_3), nitrous oxide (N_2O), and dinitrogen tetroxide (N_2O_4), have been observed in CAP [39–42]. Positively charged ions such as N_2^+ and electrons are also generated by CAP [3, 39].

Direct treatment is performed when cells are in culture conditions or in an environment similar to culture conditions. When a direct treatment was performed, cells were always covered by a thin layer (<1 mm) of cell culture medium or other biologically adaptable solutions [43]. The potential biological effects from physical factors such as heat, ultraviolet, as well as the plasma's electromagnetic field may be blocked by this aqueous layer, particularly when this layer is adequately thick. That may explain why the physical effect had not been found in nearly all previous studies. In contrast, such a thin layer of biologically adaptable solution is able to facilitate the transition of short-lived reactive species in gas phase to long-lived reactive species in liquid phase [44–48]. Without this aqueous interface, many chemical reactions may be interfered with or even blocked. A liquid interface of culture medium or PBS also contains buffering chemicals such as phosphate monobasic ($H_2PO_4^-$), phosphate dibasic (HPO_4^{2-}), and bicarbonate (HCO_3^-) that can prevent short-duration CAP treatment from causing drastic changes in pH [49, 50]. In contrast, the absence of buffer in aqueous solutions will result in a sharp pH change in the CAP-treated water by the direct treatment or underwater discharge process [51].

CAP also generates several physical factors, mainly electromagnetic emissions. Different wavelengths in the EM spectrum, including thermal (infrared), electromagnetic, and ultraviolet emissions, exert diverse potential effects on cells. The biological effect of these physical factors has rarely been observed over the past decade. As a result, these physical factors have previously been regarded as a minor or even neglected factor when considering the anti-cancer performance or mechanism. More recently, the physically-based treatment and strong anti-cancer efficacy has been reported. Chapter 6 systematiclly introduced the research status of the physically-based treatment.

Direct and indirect treatment share many similarities [52]. First, nearly all cellular damage can be observed in both direct and indirect treatment. Many distinct mechanisms and cellular responses have been observed in both treatment approaches, including the rise of intracellular ROS, DNA, mitochondrial, endoplasmic reticulum, and cytoplasmic membrane damage (figure 4.2) [53–57]. Besides, the sensitivity of cancer cell lines to direct and indirect treatment is also quite similar.

However, practically, direct and indirect treatment represent two quite different approaches of using CAP in cancer therapy. In the clinic, the former fits within a surgery-like or surgery-assistant modality, while the latter represents a pharmacological approach. Though the biological effect of CAP on cancer cells has been studied for nearly two decades, what we learned so far is mainly focused on the effects caused by long-lived reactive species, which led to a limited understanding of CAP's uniqueness (figure 4.3). In our opinion, CAP treatment is different from a simple reactive species treatment. The uniqueness of CAP relies on its capacity to generate both chemically reactive species and physical factors to cause the reported biological effects. The indirect treatment may solely depend on the long-lived reactive species, which can be easily obtained by other methods. The perspectives based on long-lived reactive species have contributed significantly to our understanding of the interaction between CAP and cells. But the promise of CAP lies in discovering new biological effects of bulk plasma; by shifting efforts to focus on understanding direct treatment we can better understand the uniqueness of CAP as an anti-cancer modality.

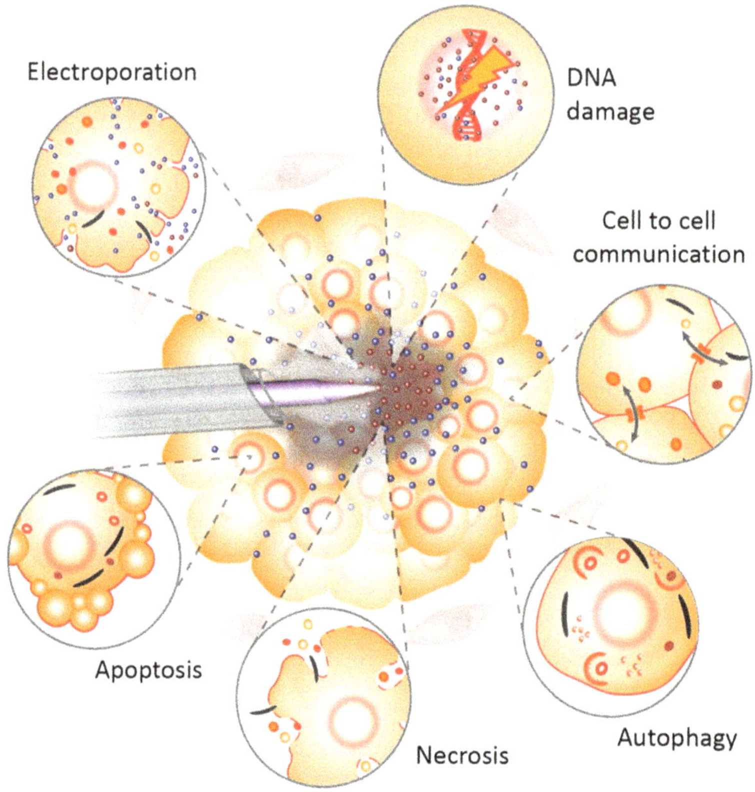

Figure 4.2. Common cellular damages were observed in both direct and indirect treatment *in vitro*. Reproduced from [23], CC BY 4.0.

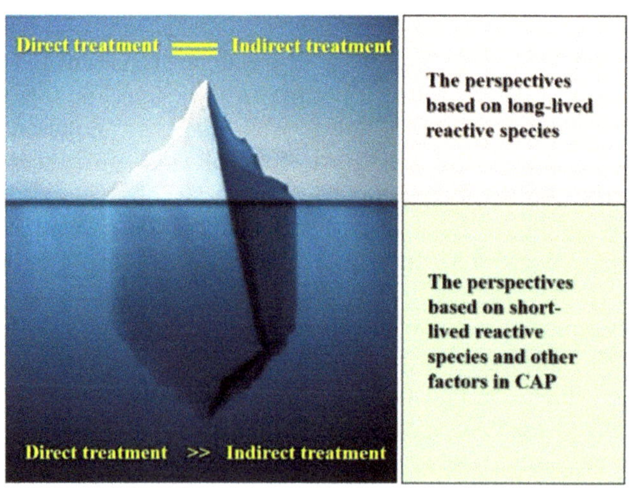

Figure 4.3. The tip of the iceberg. Direct treatment or the direct impact of bulk plasma on cells involves more CAP-related basic problems and unique features. Plasma medicine should focus more on discovering novel cellular responses to bulk plasma, which is the foundation of plasma medicine's uniqueness. Reproduced from [58] with permission from the Royal Society of Chemistry.

4.3 Anti-cancer mechanism

4.3.1 General pictures

To date, the promising anti-cancer performance of CAP treatment has been extensively demonstrated in many cancer types and even more cell lines [59, 60]. Hundreds of articles and reviews have been published [23, 61, 62]. Now, a basic question arises: what exactly is CAP doing to cells? Obviously, dissecting the cellular responses will provide the best answer to this critical question. We note that several basic cellular responses have been consistently observed and list them in table 4.1. These basic cellular responses have built the foundation we have today to understand the anti-cancer effect in a CAP treatment.

Discovering novel and unique cellular responses should be a priority in future plasma medicine. For example, we found that NO_2-gradually accumulated in extracellular space after treatment (figure 4.4) while cancer cells without CAP treatment did not generate NO_2^- in the culture process. Breast cancer cell line MDA-MB-231 and glioblastoma cell line U87MG gradually generated observable NO_2^- in the medium 6–12 h post treatment [76]. Two days after treating U87MG cells with CAP, NO_2^- generation could be as much as 10-fold higher than that measured in the CAP-treated (3.16 kV) medium without cells [76]. Currently, there is little knowledge of its underlying mechanism, but understanding this interesting phenomenon will help in building the full spectrum of CAPs effect on cancer cells.

Table 4.1. Some basic cancer cellular responses to CAP. Reproduced from [63].

References	Cancer cellular response	Years
[1]	Apoptosis	2004
[2]	Growth inhibition	2007
[9]	Cytoskeletal damage	2009
[64]	Selective cell death	2010
[65]	Cell cycle arrest	2010
[66]	Nuclear and DNA damage	2010
[66]	Mitochondrial damage	2010
[67]	Rise of intracellular ROS	2011
[13]	Chemically based sensitization to drugs	2013
[19]	Selective rise of intracellular ROS	2013
[68]	Senescence	2013
[69]	Immunogenic cell death	2015
[70]	Cell-based H_2O_2 generation	2017
[71]	Autophagy-associated cell death	2017
[72]	Activation phenomena	2018
[73]	Physically triggered necrosis	2020
[74]	Pyroptosis	2020
[75]	Physically based sensitization to drugs	2021

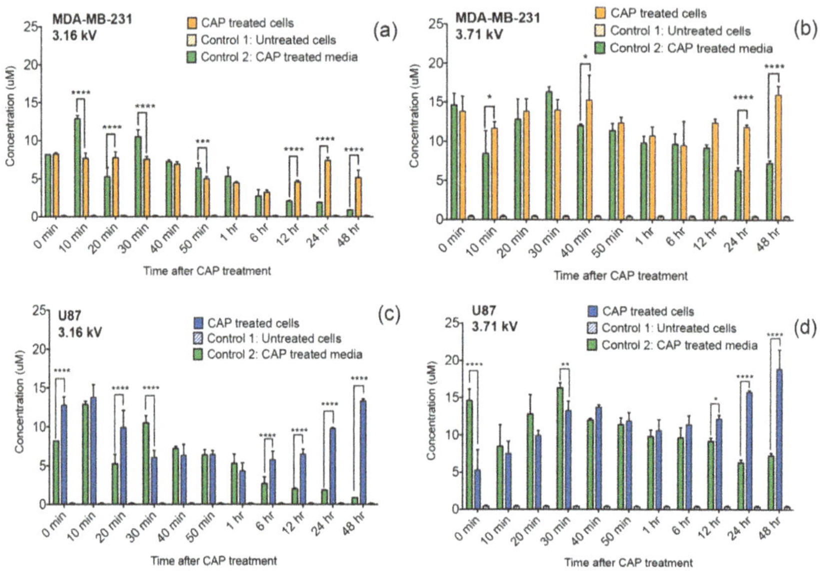

Figure 4.4. Accumulation of NO_2^- in cancer cells' extracellular space [77]. Breast cancer MDA-MB-231 cells (a and b) and glioblastoma U87MG cells (c and d) were treated with an APPJ output voltage of 3.16 or 3.71 kV for 60 s and were monitored for 48 h. Reprinted with permission from [77], copyright (2018) American Chemical Society.

In plasma medicine, cellular responses are mainly attributed to the long-term effects of CAP, which typically are seen hours or even several days after treatment. Current research associates the long-term effects of CAP with several cell damage mechanisms. Among them, the significant rise of intracellular ROS is the most widely observed cellular response [8, 19, 54, 67, 78–84]. Nonetheless, a few studies have reported ROS-independent cytotoxicity. These conclusions were mainly based on the observation that pretreating cells with N-acetylcysteine (NAC), an antioxidant, at a specific concentration does not significantly change the cytotoxicity [83, 85]. These conclusions are still questionable. Because an adequately high NAC may still counteract the cytotoxicity.

As a framework, some general conclusions could be given here (figure 4.5). Reactive species mainly ROS damage cellular function through an overturn of intracellular redox balance, causing the strong rise of intracellular ROS in complex pathways [67, 79, 81, 86]. The boost of intracellular ROS further damages a series of important cellular components, including the cellular membrane, nucleus, mitochondria, and endoplasmic reticulum. Finally, cell death by apoptosis, necrosis, and autophagy-associated cell death is triggered [41, 66, 68, 71, 87–89].

4.3.2 Rise of intracellular ROS

The rise of intracellular ROS has been widely observed in many CAP-treated cancer cell lines [19, 67]. Notably, CAP treatment cannot kill cancer cells when these cells have been pretreated with intracellular ROS scavengers such as NAC, rotenone,

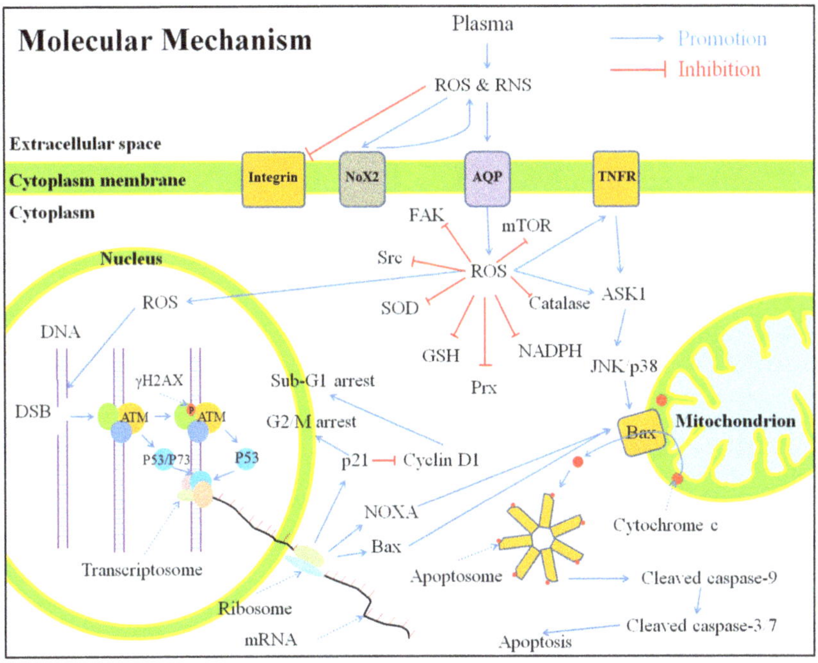

Figure 4.5. ROS-focused anti-cancer pathways. Abbreviations: ROS: reactive oxygen species, RNS: reactive nitrogen species, Nox: NADPH oxidases, AQP: aquaporins, TNFR: tumor necrosis factor receptor, FAK, focal adhesion kinase, Src: Src kinase, SOD: superoxide dismutase, GSH: glutathione, Prx: peroxiredoxin, NADPH: reduced nicotinamide adenine dinucleotide phosphate, mTOR: mechanistic target of rapamycin, DNA: deoxyribonucleic acid, DSB: double-strand break, ATM: ataxia telangiectasia mutated, mRNA: messenger ribonucleic acid, ASK: apoptosis signal-regulating kinase, JNK: c-Jun N-terminal kinase. Rperoduced from [38]. CC BY 3.0.

D-mannitol, diphenyleneiodonium (DPI), as well as apocynin [8, 14, 16, 19, 67, 80, 84, 90–93]. This evidence strongly supports that the intracellular ROS rise is one of the key factors to understand the anti-cancer mechanism of CAP treatment. Nevertheless, its underlying molecular mechanisms are not fully elucidated.

Even the chemical essence of the intracellular ROS rise is still mostly unknown because ROS assays used in plasma medicine nearly all use 2,7-dichlorodihydrofluorescein diacetate (H_2DCFDA), an oxidant-sensitive fluorescent dye, but a non-specific ROS probe [94, 95]. One cannot pinpoint exactly which reactive species play a role and how big that role is since H_2O_2, O_2^-, $OH^\bullet$, and $ONOO^-$ may all contribute to the increased intracellular ROS [50, 67, 81, 96, 97]. Basically, intracellular ROS rise may be due to the diffusion of extracellular CAP-originated ROS or RNS [8, 19, 79, 98, 99]. This explanation is consistent with the observation that the rise of intracellular ROS will be observed just several minutes after CAP treatment [79, 80]. Since most reactive species are either charged or polar molecules, specific membrane channels or transporters in the cytoplasmic membrane are necessary [100, 101]. Among CAP-generated reactive species, NO may freely cross the phospholipid membrane [79]. In addition, $ONOO^-$ can freely cross phospholipid

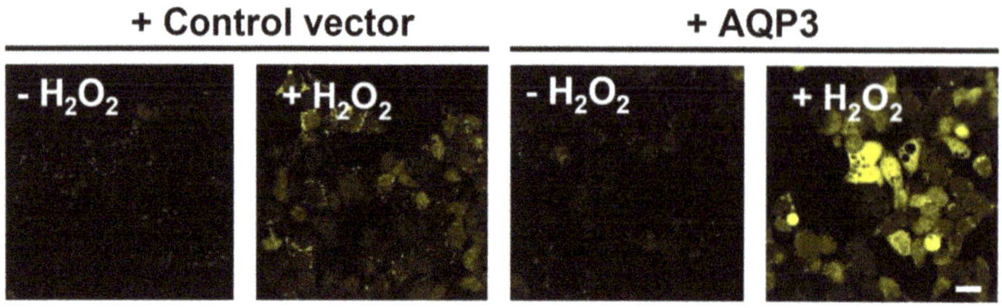

Figure 4.6. AQP3 increases H_2O_2 uptake in HEK 293 cells [107]. Molecular imaging was performed using peroxy yellow 1 methyl-ester (PY1-ME), a specific chemoselective fluorescent indicator for H_2O_2. HEK 293 cells were transfected with either a control vector or AQP3 and monomeric red fluorescent protein (mRFP) as a transfection marker and measured by confocal microscopy. While scale bar: 20 μm. Reproduced from [107]. CC BY 4.0.

membranes with a permeability coefficient of about 400 times greater than that of O_2^-, although some unknown anion channels may also be involved in its transmembrane diffusion [102, 103].

Aquaporins (AQPs), were first discovered as a specific water channel decades ago [104]. Recently, the role of AQPs in the transmembrane diffusion of other small molecules such as H_2O_2, CO_2, NO, NH_3, urea, and even glycerol was confirmed [105]. Within the AQPs family, at least AQP1, 3, 8, and 9 have been demonstrated as the channels for H_2O_2 transmembrane diffusion [106–109]. AQP1 can facilitate the passive diffusion of NO across cell membranes [110]. In glioblastoma (U87MG) cells, AQP8 plays a role in the transmembrane diffusion of H_2O_2 after CAP treatment. Silencing the expression of AQP8 could inhibit the cytotoxicity of CAP treatment [111]. In addition, H_2O_2 specific chemical staining in colon adenocarcinoma HT29 cell line, mammalian human embryonic kidney cell line (HEK 293), and cervical cancer HeLa cell line revealed that the H_2O_2 uptake was drastically increased with higher expression of AQP3 on their cellular membrane (figure 4.6) [107]. A similar trend was also observed in other eukaryotes. Yeast cells expressing AQP1 or AQP8 had a much stronger resistance to H_2O_2 compared with the yeast cells without AQP1 or AQP8 [108]. A molecular dynamics simulation further demonstrated that free energy barrier of H_2O_2 across AQP1 was lower than that of palmitoyl-oleoyl-phosphatidylcholine phospholipid bilayer, indicating that the delivery of H_2O_2 into the cell through AQP is theoretically supported [112].

The rise of intracellular ROS may also be due to the diffusion of extracellular RNS. In particular, NO can trigger intracellular ROS pathways. For example, NO caused the rise of intracellular ROS such as H_2O_2 and RNS such as NO in HeLa cells exposed to CAP-treated NO-containing water [99]. As mentioned above, the rise of intracellular RNS may be simply due to the transmembrane diffusion of NO. Besides, the intracellular ROS rise may result from the NO-dependent blockage of the electron transfer chain located in the inner membrane of mitochondrion [98].

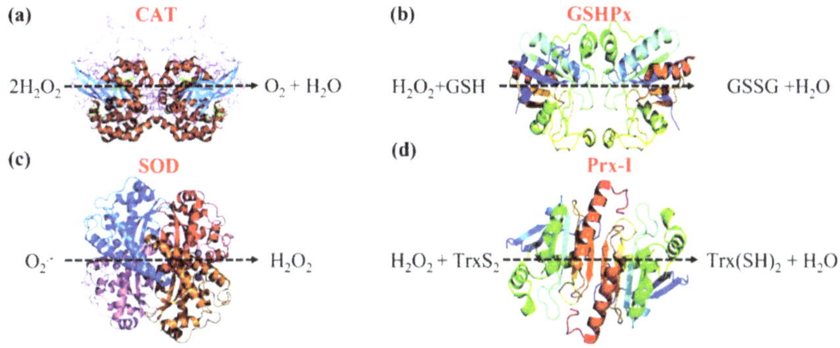

Figure 4.7. Structures and functions of four typical enzymatic antioxidants. (a) CAT; (b) GSHPx; (c) SOD; (d) Prx-1. Reproduced from [119]. CC BY 3.0.

Certain intracellular pathways and reactions may also contribute to the intracellular ROS rise. One pathway may involve overturning the redox balance by compromising the intracellular antioxidant system. Naturally, the redox balance is achieved by resisting the change of intracellular oxidative stress through a complex cellular antioxidant system, which includes small molecules such as glutathione (GSH), reduced nicotinamide adenine dinucleotide phosphate (NADPH), as well as ROS-scavenging enzymes such as superoxide dismutase (SOD), catalase (CAT), thioredoxin reductase, peroxiredoxin (Prx), glutathione peroxidases (GSHPx)/reductases (GSHR) (figure 4.7) [113–116]. The decreased NADPH/$NADP^+$ ratio and GSH/GSSG ratio have been observed in CAP-treated cancer cells [3, 117]. Similarly, the decrease of Prx, SOD, as well as CAT has also been observed after treatment [3, 16, 118]. Decreasing the expression of Cu, Zn-SOD or Mn-SOD also increased the cell death of HeLa cancer cells after treatment, suggesting a relationship between ROS and the intricate pathways of the antioxidant system [84].

In addition, the rise of intracellular ROS may also be due to secondary ROS originated from membrane oxidases [94]. The NADPH oxidase (Nox) on the cytoplasmic membrane was found to cause the rise of intracellular ROS and the decreased viability of colorectal cancer cells after a CAP treatment [94]. The CAP-originated reactive species may activate the expression of *Nox2* in cancer cells, but they do not directly enter the cytoplasm [94]. Nox may generate secondary ROS such as superoxide, which further inhibits the cell growth [94]. However, the transmembrane diffusion of these secondary ROS such as H_2O_2 also relies on AQP on cellular membrane [107, 120].

Recently, molecular simulation (MD) provided another perspective to understand transmembrane diffusion of reactive species. It was found that transient pores (~15 Å) could be naturally formed on phospholipid bilayer (PLB) when all phospholipids were oxidized (figure 4.8) [121]. These transient pores on the cellular membrane provide routes for reactive species to quickly enter the cytosol. Further studies found that RNS (NO, NO_2, N_2O_4) and O_3 may permeate more easily

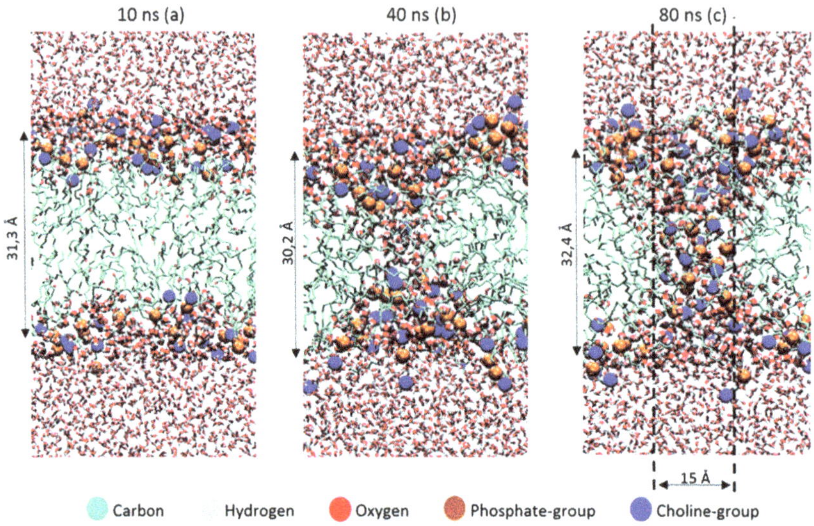

Figure 4.8. Snapshots of transient pore formation on PLB based on MD simulation, after 10, 40, and 80 ns in a model system with 100% oxidation. Reproduced from [124]. CC BY 3.0.

through the oxidized PLB than hydrophilic ROS (OH, HO_2, H_2O_2) [122]. More interestingly, cell membranes consisting of combined nitro-oxidized lipid products may inhibit the transient pore formation in the presence of oxidized lipids [123].

4.3.3 Other cellular damage

The damage on the cellular membrane and membrane proteins are another long-term cellular response to CAP [125]. The cellular membrane is the first barrier between the cells and the CAP-originated reactive species. Just shortly after an adequately high dose treatment, many cancer cells experienced a noticeable morphological change from expanded to highly contracted shapes [9, 27, 66]. The horizontal polarization and cytoplasmic protrusions of cells were lost in these contractive shapes [126, 127]. Cells are shaped by the cytoskeleton system. The altered cell shape was indeed accompanied by a fundamental architecture change in the cytoskeleton particularly with respect to F-actin [9, 126]. In addition, CAP also decreased the expression of the components of the extracellular matrices such as integrin [94,128], Src kinase (Src) [126], focal adhesion kinase (FAK) [9], as well as Rho family [126]. These changes may explain the decreased migration rate and the increased detachment rate of cancer cells upon treatment [7, 128]. Moreover, the degradation and dysfunction of the epidermal growth factor receptor (EGFR) occurred on CAP-treated EGFR-overexpressing oral squamous cell carcinoma but not in the normal cell line [129]. Such damage on EGFR may be induced by the intracellular ROS rise [129].

DNA damage has been commonly observed as an early-stage event after CAP treatment [59, 130]. The double strand break (DSB) is the main damage caused by CAP [8, 42]. An important marker of DSB is a specific phosphorylation on serine

139 on H_2AX histone (γ-H_2AX), which has been widely observed after treatment [8, 39, 42]. The serine 139 on H_2AX is phosphorylated by the ataxia telangiectasia mutated (ATM) recruited to the DSB site with other DNA damage response complexes [103, 131]. The enhanced expression of ATM was also observed in the CAP-treated oral cavity squamous cell carcinoma cells [132]. The activated monomeric ATM can also phosphorylate p53 and other substrates in the DNA damage response complexes localized at DSB sites [132]. Phosphorylation of p53 (p-p53) was observed in the CAP-treated oral cavity squamous cell carcinoma cells and melanoma cells [68, 132]. In addition, the increased p-p53 expression in the CAP-treated mouse melanoma cells (B16F10) was associated with γ-H_2AX expression [66], which provided clues to understand the chronological order of these pathways. Moreover, ATM may also phosphorylate p73, which further activates the expression of pro-apoptotic factors [103]. The significantly enhanced expression of p73 in the CAP-treated melanoma (Me1007) cells was also observed [133], while knockout of p73 gene in Me1007 cells noticeably reduced CAP's anti-cancer performance [133].

Cell cycle arrest is another common cellular response to CAP treatment. The phosphorylation of p53 is a necessary step for a series of downstream cell cycle arrest pathways [103, 131]. Activation of p21 expression by p53 was observed in a diverse group of CAP-treated cancer cell lines (figure 4.9) [68, 132]. This is significant because activated p21 is known to trigger the different types of cell cycle arrests [134]. In the CAP-treated oral cavity squamous cell carcinoma cells, p21 inhibited the function of cyclin D1, causing noticeable sub-G1 arrest [132]. In the CAP-treated melanoma cells, p21 caused a G2/M arrest by an unknown pathway [68]. G2/M cell cycle arrest was also observed in several other treated cancer cell lines [13, 135, 136].

4.3.4 Cell death

Cell death is a final fate of many CAP-treated cancer cells. The main death mechanism of CAP-treated cancer cells is apoptosis, though necrosis and autophagy-associated cell death have also been observed [59, 71, 130, 137, 138]. Apoptosis is caused by reactive species particularly ROS [44, 56, 57, 90]. Both extra-/intra-cellular ROS scavengers, such as NAC, and apoptosis inhibitors, such as zVAD, effectively inhibit apoptosis in CAP-treated mammalian cells [6, 16, 67, 81, 94, 95, 133]. Apoptotic characterization is mainly based on flow cytometric data [2, 18, 117, 139, 140], accompanied by analysis of western blots, polymerase chain reactions (PCR), and other biochemical assays [5, 40, 53, 54, 95, 141, 142].

As a specific form of programmed cell death, apoptosis plays critical roles in many human pathologies, including cancer, autoimmune diseases, as well as neurodegenerative disorders [143, 144]. Typically, apoptosis in mammalian cells is characterized by a series of chronological features, including cell shrinkage, cytoplasmic membrane blebbing, chromatin condensation, nuclear membrane breakdown, and ultimate formation of apoptotic bodies [145]. Apoptotic bodies have complete cellular membranes and will be cleared by phagocytes *in vivo*, avoiding triggering

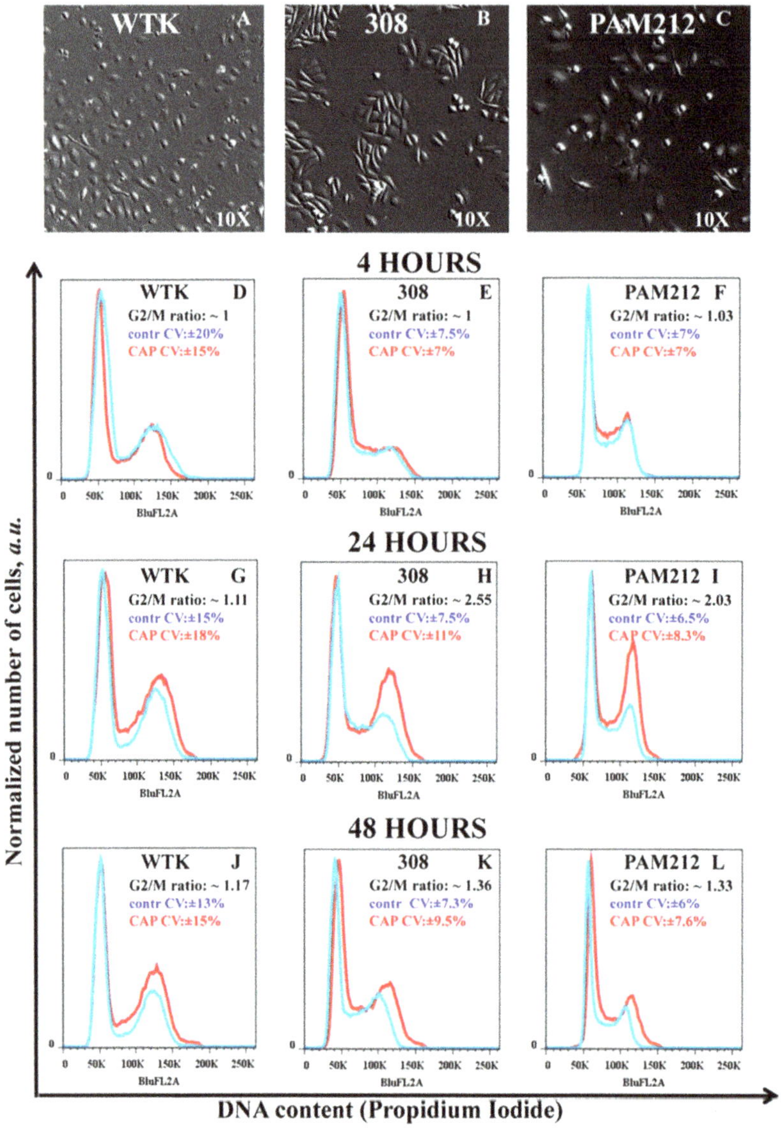

Figure 4.9. G2/M cell cycle arrest [135]. (A–C) Bright-field images of wild type keratinocytes (WTK), epidermal papilloma (308 cells) and epidermal carcinoma (PAM212 cells) cells. (D–L) show cell cycle flow cytometry studies: Propidium Iodide content (horizontal axis) and normalized number of cells (vertical axis) were shown. The controls were shown in blue and the cells after treatment of 60 s were in red. The ratio of the number of cells (treated to untreated) in G2/M-phase with coefficients of variation (CV, in percents) was shown in the right top corner of each figure. Measurements were performed 4, 24, and 48 h after treatment. Reprinted from [135], copyright (2012), with permission from Springer Nature.

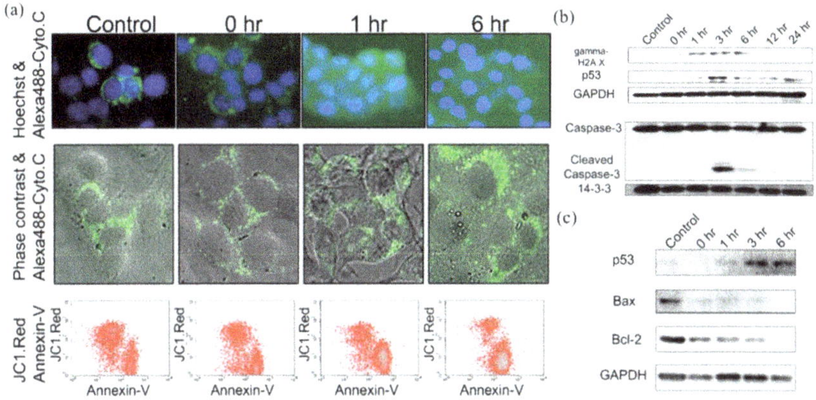

Figure 4.10. A typical apoptotic observation in the CAP-treated melanoma B16F10 cells [66]. (a) Cytochrome c releasing by immunocytochemistry and fluorescence-activated cell sorting analysis for mitochondria membrane potential change. (b) Western blot of the phosphorylation on γ-H2AX, p53, and caspase-3. (c) Western blot of BAX and BCL-2. Reprinted from [66], with the permission of AIP Publishing.

inflammatory responses [145, 146]. The basic biochemical mechanism of apoptosis has been extensively investigated over the past four decades and the basic conclusions were systematically explained in many references [147, 148].

Generally, the CAP-triggered apoptosis strictly follows the well-understood apoptotic events and pathways. CAP-triggered cancer cells' apoptosis mainly followed the caspase-dependent apoptosis pathways [2, 41, 49, 68, 81, 94, 133, 149–153], though a few studies reported the caspase-independent pathways [57, 154]. Most apoptotic pathways in the CAP-treated cancer cells are the mitochondrial pathways triggered by DNA, mitochondrial, as well as endoplasmic reticulum damage (figure 4.10) [8, 13, 42, 57, 59, 130]. The release of cytochrome c into the cytosol, the release of calcium, the expression of NOXA, Bax, caspase-3/7, caspase-9, the loss of mitochondrial transmembrane potential, DNA fragmentation, as well as the cleavage of poly ADP-ribose polymerase (PARP), have been observed in CAP-treated cancer cells [13, 21, 32, 41, 49, 53, 57, 66–68, 87, 88, 91, 117, 149, 155, 156]. In addition, the tumor necrosis factor receptor (TNFR)-based apoptotic pathway was also activated in CAP-treated head and neck cancer cells [149].

Autophagy-associated cell death is as a novel mode of cell death [157]. In biology, autophagy was first discovered as a survival mechanism employed by yeast cells exposed to sublethal stresses, such as nutrient deprivation or the lack of extracellular growth factors [157–159]. Autophagic cells can survive by digesting their own organelles and macromolecules and recycling these damaged organelles or macromolecular components [157, 160]. However, if prolonged, autophagic cells can ultimately digest all available substrates and die, which is an autophagy-associated cell death [157, 161]. Such a cell death is characterized by the formation of abundant autophagosomes (AP) in dying cells [162]. Oxidative stress condition or ROS exposure is a typical mechanism to trigger autophagy-associated cell death [163].

Recently, autophagy in the CAP-treated cancer cells has been demonstrated. The oxidative stress from CAP was the main force triggering autophagic responses in several cell lines, including, endometrial cancer cells (AMEC, HEC50), melanoma G-361, and mesothelioma cells EM2 [71, 89, 164]. Specifically, the synergistic use of CAP and silymarin nanoemulsion (SN) triggers autophagy in G-361 cell line by activating PI3K/mTOR and epidermal growth factor receptor (EGFR) pathways [89]. For endometrial cancer cells AMEC and HEC50, CAP-originated long-lived reactive species inactivated the mTOR pathway [164]. Alongside its discovery, it was found that autophagic cell death could be partially inhibited by an autophagy inhibitor MHY1485 [164].

Here, we introduce a representative example about the autophagic cell death of the CAP-treated malignant mesothelioma cells. The autophagic response of malignant mesothelioma cells was induced by CAP's oxidative stress through a reaction involving cellular iron (figure 4.11(a)). CAP also caused a rapid nuclear translocation of a transcription factor, transcription factor EB (TFEB), which further regulated lysosomal biogenesis and autophagy [71]. Direct transmission electron microscopy (TEM) imaging of autophagosomes (AP) and autophagolysosomes (AL) in the CAP-treated mesothelioma cells was presented in figure 4.11(b). The noticeable AP (labeled by LC3B) could be observed at 2 h after CAP treatment

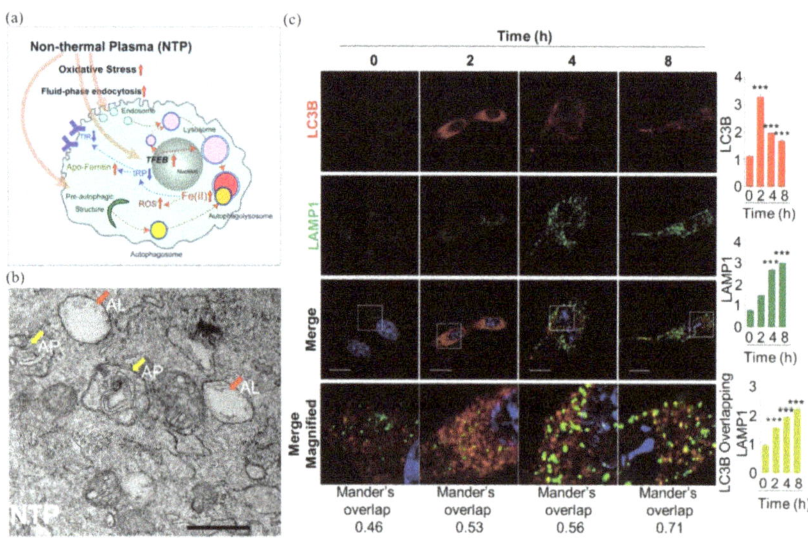

Figure 4.11. CAP induces oxidative stress response in malignant mesothelioma cells (SM2) and increases endocytosis, lysosome biogenesis, and autophagy-associated cell death. (a) Schematic autophagy mechanism. (b) TEM imaging of vesicle structures, endosome-like vesicles, and organelles in the autophagic pathway. Yellow and red arrowheads indicate AP and AL, respectively. Scale bar: 500 nm. (c) Fluorescent imaging of autophagosomes in the treated (60s) malignant mesothelioma cells. Cells were stained by DAPI (DNA, blue), LC3B (red), and LAMP1 (green), respectively. LC3B suggested the presence of autophagosomes, while the colocalization (yellow overlay fluorescence) of LC3B and LAMP1 suggested the autophagolysosome formation. Scale bar: 20 μm. Reprinted from [71], copyright (2017), with permission from Elsevier.

(figure 4.11(c)). AL, labeled by LC3B and lysosomal-associated membrane protein 1 (LAMP1), were observed at 4 h after CAP treatment [63].

4.3.5 Chemical mechanism: reactive species

Reactive species have been regarded as the basis of CAP's anti-cancer effect. Most basic cellular respones are due to the impact of reactive species, though a few of them are due to physical impact, which is systematically discussed in chapter 6. Typical ROS include hydroxyl radical ($^{\bullet}OH$), singlet oxygen ($^{1}O_2$), superoxide anion (O_2^-), and hydrogen peroxide (H_2O_2) [151]. Typical RNS include nitrite (NO_2^-), nitrate (NO_3^-), nitric oxide (NO), and peroxynitrite ($ONOO^-$) [79, 165]. Many reactive species are naturally toxic to cancer cells. For example, a typical short-lived reactive species, $OH^{\bullet}$, is highly reactive to many biomolecules, including proteins, phospholipid, DNA, and RNA [166]. Another short-lived reactive species, $ONOO^-$, can damage antioxidant enzymes and other critical biomolecules [167, 168]. Long-lived reactive species, for example, H_2O_2, are toxic to various cancer cell lines [169]. H_2O_2 can cause oxidation to proteins and DNA [170].

The main reactive species in CAP can be divided into two categories: long-lived and short-lived reactive species. The sharp difference in reactive species' lifetime length significantly affects their abilities and approaches to impact cells (table 4.2). Short-lived reactive species such as $OH^{\bullet}$, $^{1}O_2$, $ONOO^-$, and O_2^-, interact with cells at the plasma interface over a short-time length scale (ns, μs, and ms) [171–173]. Long-lived reactive species such as H_2O_2, NO, NO_2^-, NO_3^-, as well as O_3, may interact with the cells at a site far from the contacting point of plasma after diffusing over a long-time length scale [174].

Several long-lived ROS such as H_2O_2, and long-lived RNS such as NO_2^-, NO_3^-, and $ONOO^-$ have been widely detected in CAP-activated solutions (PAS) [44, 49, 81, 150, 151, 165, 181]. In addition, many studies also found NO in plasma activated medium (PAM) [79, 81, 150]. Many components in the culture medium particularly amino acids or proteins are modified upon CAP treatment [181, 182]. Among the 20 amino acids, cysteine shows the strongest reactivity towards the plasma-originated reactive species dissolved in PSM [44]. Initially, H_2O_2 was widely

Table 4.2. Half-life of some reactive species *in vitro*.

Species	Half-life	Reference
Hydroxyl radical ($OH^{\bullet}$)	10^{-9} s	[175]
Superoxide radical ($O_2^{\bullet-}$)	10^{-3} s	[176]
Hydrogen peroxide (H_2O_2)	10^{-5} s	[173]
Hydrogen peroxide (H_2O_2)	Days, weeks	[177]
Singlet oxygen ($^{1}O_2$)	10^{-6} s	[172]
Nitric oxide radical ($NO^{\bullet}$)	Hours	[178]
Peroxynitrite ($ONOO^-$)	10^{-3} s	[179]
Nitrite (NO_2^-)	Months	[180]
Nitrate (NO_3^-)	Months	[77]

regarded as the main anti-cancer factor causing cell death *in vitro* [3, 44, 50, 55, 57, 81, 96, 177, 183–185]. Extracellular H_2O_2 scavengers such as catalase, cysteine, sodium pyruvate, as well as some cysteine derivatives such as N-Acetyl-Cysteine (NAC) will significantly weaken the efficacy of CAP treatment [16, 44, 85, 96, 150]. Besides, for some H_2O_2-resistant cell lines, the cytotoxicity of CAP treatment will be much stronger than the H_2O_2-treatment with the same H_2O_2 concentration. Other CAP-originated products, such as $^1O_2$, O_2^-, $OH^•$, as well as $ONOO^-$ may also inhibit cell growth [91, 186]. For example, $ONOO^-$ was regarded as the key factor to induce the apoptosis of the nanosecond pulsed N_2/O_2 plasma-treated myeloma cells [187].

In plasma medicine, a basic concern is the role of plasma during treatment. Nearly all previous studies provided a similar but simplified picture, that a CAP treatment was just a process generating many reactive species in the extracellular environment [45]. Evidence for such a conclusion is seen when extracellar reactive species are immediately removed by replacing ’the medium surrounding CAP treated cells and it completely inhibits CAP’s cytotoxicity [188]. The presence of extracellular ROS scavengers such as cysteine, pyruvate, as well as nordihydro-guaiarectic acid (NDGA) can counteract CAP’s cytotoxicity as well [19, 189].

Our experimental evidence demonstrated that such a picture was not complete. Based on recent literature, CAP plays at least three important roles. The first role of CAP treatment, as discussed above, is providing abundant reactive species in the extracellular environment. Reactive species are required for CAP to exert cytotoxicity on cancer cells. To affect the growth of cancer cells or to cause the cell death, these reactive species need to accumulate during a sufficiently long treatment, lasting minutes or even longer [44, 56, 71, 190]. Indirect treatment has similar effects on cancer cells but we must consider the potential role of new products formed in PAS [44, 177, 181]. However, such an understanding cannot explain the observation that a stronger anti-cancer effect is seen in direct treatment on cancer cells versus indirect treatment using PAS (figure 4.12). Two more unique roles of CAP in cancer killing are activation phenomena and cell-based H_2O_2 generation, which will be discussed in following sections.

4.3.6 Activation phenomena

The activation of cancer cells may be a unique feature of CAP treatment [192]. CAP-triggered activation will not cause killing effect, but activation drastically decreases the threshold of these cancer cells to the cytotoxicity of reactive species, particularly ROS (figure 4.13) [192]. For example, 22 μM H_2O_2 causes negligible growth inhibition effect on the pancreatic cancer cells without the activation of CAP. In contrast, the same H_2O_2 treatment can cause as high as 50% decrease in the cell viability of the CAP-activated pancreatic cancer cells. In previous studies, RNS such as NO_2^- and NO_3^- have been regarded as the safe chemicals for cancer cells even with a concentration as high as 900 μM [190]. However, it is found that just 50 μM NO_2^- can cause noticeable growth inhibition of pancreatic cancer cells, if cells are treated and activated by CAP.

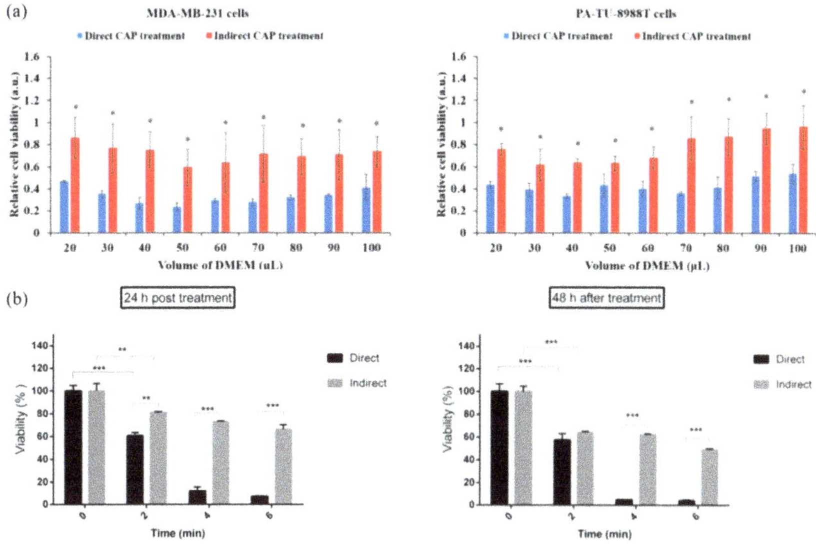

Figure 4.12. Direct versus indirect treatment. (a) Comparative studies on breast cancer cell line MDA-MB-231 and pancreatic cancer cell line PA-TU-8988T [188]. Reproduced from [188]. CC BY 4.0. (b) Comparative studies on melanoma cell line B16F10 [191]. Reproduced from [191]. CC BY 4.0.

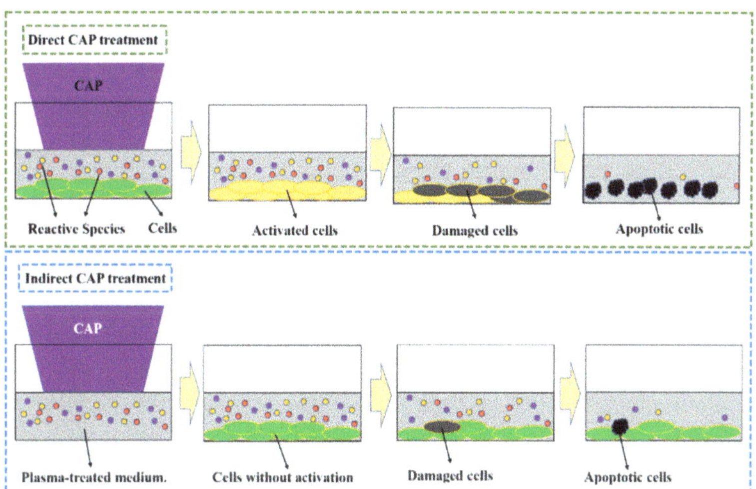

Figure 4.13. A schematic illustration of the CAP-based cell activation phenomenon during direct treatment. When reactive species in the CAP-treated medium is the same, the activated cells are much more sensitive to reactive species than the indirect treatment without activation phenomenon. Reproduced from [192]. CC BY 4.0.

The CAP-treated cells will enter an activation state. We first observed the activation state of the CAP-treated pancreatic cancer cells, which drastically advanced our basic understanding. CAP-activated medium (PAM) causes a much stronger killing effect on the CAP-activated pancreatic adenocarcinoma cells

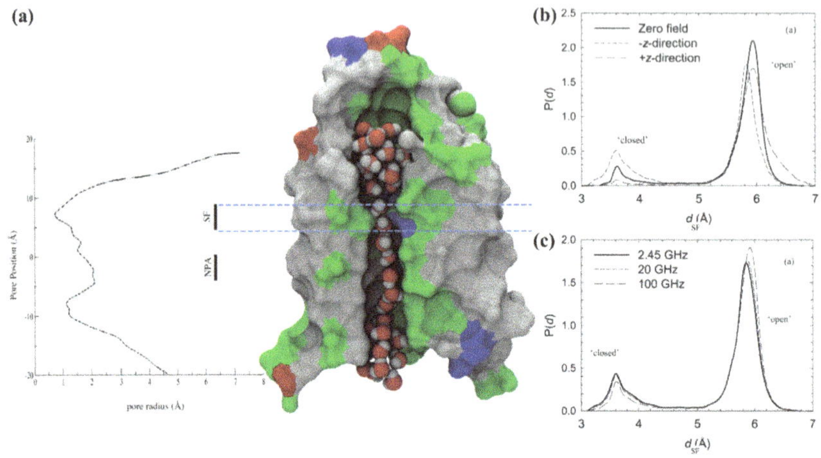

Figure 4.14. Externally applied electric field affects AQP4 function by affecting its pore radius. (a) Average pore radius profile (left) along the axis for the four pores calculated with the program HOLE for zero-field simulation. A MD snapshot (right) matching the pore radius profile. Reprinted from [193], with the permission of AIP Publishing. (b) Normalized probability distribution of SF diameter (d_{SF}) averaged over four pores and sampled during zero-field conditions and for the static electric fields applied along z-directions [194]. (c) Normalized probability distribution of d_{SF} averaged over four pores and sampled during 2.45, 20, and 100 GHz conditions. Reprinted from [194], with the permission of AIP Publishing.

compared with the same cell line without activation. In other words, the cytotoxicity of the CAP-originated reactive species is amplified through activation process. Therefore, a direct treatment will cause stronger cytotoxicity on cancer cells compared with indirect treatment. This has been observed across numerous studies [70, 191]. Activation phenomenon may be a fundamental difference between CAP treatment and other chemical treatments including the indirect CAP treatment, although the underlying mechanism has yet to be identified.

Activation phenomena has not been observed in indirect CAP treatment. Thus, physical factors or short-lived reactive species in a direct treatment may trigger such an activation phenomenon. For example, an external DC or AC electric field may affect the opening and closing of specific channel proteins. Aquaporins are important channcls for ROS particularly H_2O_2 through thc cellular mcmbranc [193, 194]. The pore radius size at selectivity filter (SF) region is an important parameter to reflect the conductivity of AQP to H_2O and H_2O_2 (figure 4.14(a)). Both DC and AC field altered water dynamics within the luman of AQP4's pore [194]. An externally applied DC field along the channels (+z direction) faciliates to 'open' the channel in the SF ragion, and vice versa (figure 4.14(b)). When frequency of AC increased from 2.45 GHz to 100 GHz, the probability of opening in the SF region increased (figure 4.14(c)). CAP also generates a wide range of EM emission. We measured the EM spectra of a typical APPJ and found the distribution at the range of 5–30 GHz (figure 4.15). The enhanced opening of AQP will not noticeably affect the intrinsic viability of cancer cells, but under conditions of plasma treatment it may enhance the cytotoxicity of ROS by allowing faster diffusion into cells. In short,

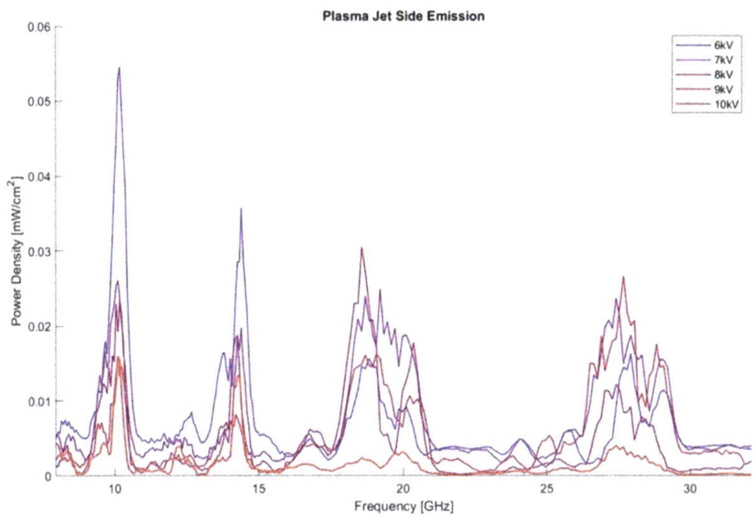

Figure 4.15. RF spectra (8–32 GHz) of an APPJ under different discharge voltages (pk–pk). Reprinted with permission from [73], copyright (2020) American Chemical Society.

the electromagnetic effect from CAP may be a candidate to explain the activation phenomenon.

A quick sensitization and a slow desensitization are two features of the CAP-based cellular activation. We found that the observable activation phenomenon started even after a treatment lasting just a few seconds. Activation phenomenon became more profound at 10–20 s after the onset of treatment. In contrast, a full de-sensitization process of the activated cancer cells took several hours after treatment. Cancer cells gradually lose their sensitivity to H_2O_2 over the initial 5 h after an activation process lasting 1 min. Such a slow de-sensitization process is even longer than the existence of ROS in the extracellular environment, which is typically 1–3 h [190]. This long duration property may give CAP-based activation a large advantage in the clinical application.

4.3.7 Cell-based H_2O_2 generation

The short-term or the instantaneous cellular responses to CAP treatment has been neglected for years. Finding instant cellular responses is a challenge. Sasaki *et al* found the transient increase of the intracellular Ca^{2+} of breast cancer cell line MCF-7 in just as short as 10 s after a CAP treatment [32]. We found another instant cellular response: strong cell-based H_2O_2 generation. CAP generates an abundance of reactive species, particularly ROS. Previously, CAP was regarded as the sole ROS source in plasma medicine. Cancer cells were just regarded as ROS consumers. A simple method was used to measure generation of H_2O_2 by treated cells [195]. Quick H_2O_2 measurement in extracellualar environments with and without cells revealed the extra H_2O_2 generation due to cells. H_2O_2 measurement was performed within 30 s after treatment, because these H_2O_2 would be quickly consumed or counteracted by cells in

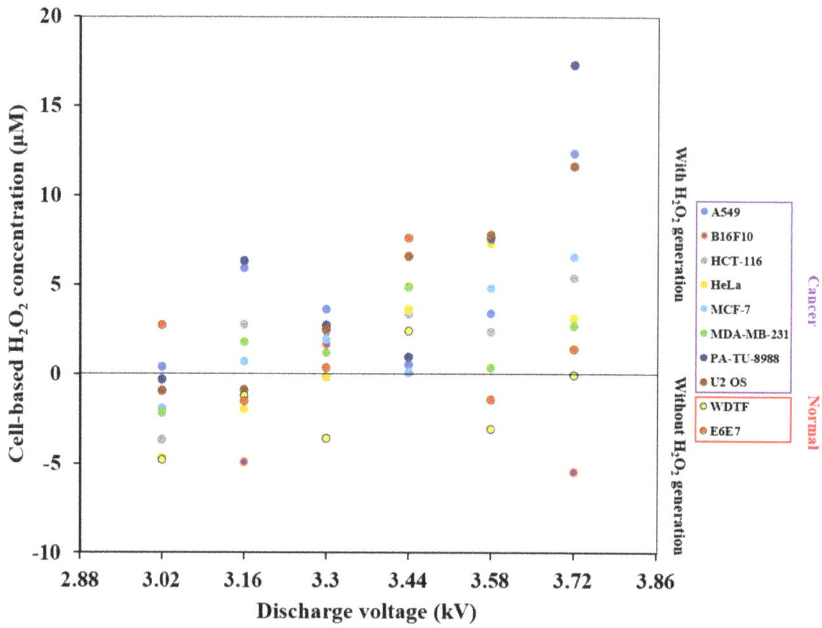

Figure 4.16. Cells-based H_2O_2 generation. Cells based H_2O_2 generation commonly existed in nearly all investigated cell lines, including eight cancer cell lines (A549, B16F10, HCT-116, HeLa, MCF-7, MDAMB-231, PA-TU-8988T, U2 OS), and two normal cell lines (WDTF and E6/E7/hTERT).

about 10 min due to an unknown mechanism [70, 76]. We observed that as many as eight cancer cells could generate strong (micromolar level) H_2O_2 during a 1 min CAP treatment (figure 4.16). Additionally, the cell-based H_2O_2 generation can also be observed in the fibroblast cell line and astrocyte cell line (E6/E7/hTERT) under certain discharge conditions. Generally, these cancer cell lines generate much more H_2O_2 than two normal cell lines, which may suggest a novel approach to realize or enhance a selective killing performance.

H_2O_2 is a critical signaling molecule in cancer cells [169]. A nanomolar (nM) level H_2O_2 production by several cancer cell lines was observed two decades ago in colon carcinoma, melanomas, neuroblastoma, and ovarian carcinoma [196]. Howcvcr, significant secretion of H_2O_2 by cancer cells has rarely been observed, which may be because a high H_2O_2 dose is selectively toxic to cancer cells [169, 197–199]. To date, CAP may be the sole physical approach to trigger a micromolar level H_2O_2 generation, which is a unique significance of this discovery. Cell-based H_2O_2 generation clearly can explain the stronger anti-cancer effect of the direct CAP treatment compared with the indirect CAP treatment under experimental conditions [70]. The discovery of cell-based H_2O_2 generation prompts us to explore more novel anti-cancer factor generation specifically due to CAP treatment, potentially providing more clues to understand the uniqueness of plasma medicine.

Cell-based H_2O_2 generation first revealed that the cells could generate and release toxic chemicals into their environment as a response to CAP treatment. When the volume of medium was adequately small, the cell-based H_2O_2 generation could be nearly as high as the H_2O_2-generated from CAP [70]. This phenomenon only occurred during the direct treatment. Therefore, short-lived reactive species such as superoxide (O_2^-) in plasma may trigger a dismutation reaction catalyzed by the extracellular superoxide dismutase (Ex-SOD) such as SOD3 on cancer cells' cytoplasmic membrane [200, 201]. SOD3 is specifically expressed in some human tumorous tissues, such as pancreatic adenocarcinoma tissues and estrogen-induced breast cancerous tissues [202–204]. Such a specific expression of SOD3 in cancer cells may cause the specific cell-based H_2O_2 generation in some cell lines such as breast cancer cell line MDA-MB-231 and pancreatic adenocarcinoma cell line PA-TU-8988T but not all cell lines such as glioblastoma cell line U87MG [70, 76].

OES spectra of APPJ under the different discharge voltages were also studied [195]. Both O and $OH^\bullet$ generation experienced a sharp increase when the discharge voltage was larger than 3.3 kV, consistent with the trend that a noticeable cell based H_2O_2 generation would occur when the discharge voltage was larger than 3.16–3.3 kV. Such an explosive increase of short-lived ROS in plasma also suggests that the cell-based H_2O_2 generation may be due to the dismutation reaction on the O_2^- provided by bulk plasma. More importantly, this work demonstrated that controlling discharge voltage can be an easy way to modulate the cell based H_2O_2 generation [70]. In general, a higher discharge voltage causes a stronger H_2O_2 generation in many cell lines.

References

[1] Kieft I E, Dvinskikh N A, Broers J L v, Slaaf D W and Stoffels E 2004 Effect of plasma needle on cultured cells *Proc. SPIE* **5483** 247–51

[2] Fridman G, Shereshevsky A, Jost M M, Brooks A D, Fridman A, Gutsol A, Vasilets V and Friedman G 2007 Floating electrode dielectric barrier discharge plasma in air promoting apoptotic behavior in melanoma skin cancer cell lines *Plasma Chem. Plasma Process.* **27** 163–76

[3] Zhao S *et al* 2013 Atmospheric pressure room temperature plasma jets facilitate oxidative and nitrative stress and lead to endoplasmic reticulum stress dependent apoptosis in HepG2 cells *PLoS One* **8** e73665

[4] Tan X, Zhao S, Lei Q, Lu X, He G and Ostrikov K 2014 Single-cell-precision microplasma-induced cancer cell apoptosis *PLoS One* **9** e101299

[5] Kim J Y, Ballato J, Foy P, Hawkins T, Wei Y, Li J and Kim S O 2011 Apoptosis of lung carcinoma cells induced by a flexible optical fiber-based cold microplasma *Biosens. Bioelectron.* **28** 333–8

[6] Plewa J M, Yousfi M, Frongia C, Eichwald O, Ducommun B, Merbahi N and Lobjois V 2014 Low-temperature plasma-induced antiproliferative effects on multi-cellular tumor spheroids *New J. Phys.* **16** 043027

[7] Wang M, Holmes B, Cheng X, Zhu W, Keidar M and Zhang L G 2013 Cold atmospheric plasma for selectively ablating metastatic breast cancer cells *PLoS One* **8** e73741

[8] Vandamme M *et al* 2012 ROS implication in a new antitumor strategy based on non-thermal plasma *Int. J. Cancer* **130** 2185–94

[9] Lee H J, Shon C H, Kim Y S, Kim S, Kim G C and Kong M G 2009 Degradation of adhesion molecules of G361 melanoma cells by a non-thermal atmospheric pressure microplasma *New J. Phys.* **11** 115026

[10] Guerrero-Preston R, Ogawa T, Uemura M, Shumulinsky G, Valle B L, Pirini F, Ravi R, Sidransky D, Keidar M and Trink B 2014 Cold atmospheric plasma treatment selectively targets head and neck squamous cell carcinoma cells *Int. J. Mol. Med.* **34** 941–6

[11] Barekzi N and Laroussi M 2012 Dose-dependent killing of leukemia cells by low-temperature plasma *J. Phys. D: Appl. Phys.* **45** 422002

[12] Ishaq M, Han Z J, Kumar S, Evans M D M and Ostrikov K 2015 Atmospheric-pressure plasma- and TRAIL-induced apoptosis in TRAIL-resistant colorectal cancer cells *Plasma Processes Polym.* **12** 574–82

[13] Köritzer J *et al* 2013 Restoration of sensitivity in chemo-resistant glioma cells by cold atmospheric plasma *PLoS One* **8** e64498

[14] Lee S, Lee H, Jeong D, Ham J, Park S, Choi E H and Kim S J 2017 Cold atmospheric plasma restores tamoxifen sensitivity in resistant MCF-7 breast cancer cell *Free Radic. Biol. Med.* **110** 280–90

[15] Park S, Kim H, Ji H W, Kim H W, Yun S H, Choi E H and Kim S J 2019 Cold atmospheric plasma restores paclitaxel sensitivity to paclitaxel-resistant breast cancer cells by reversing expression of resistance-related genes *Cancers (Basel)* **11** 2011

[16] Yang H, Lu R, Xian Y, Gan L, Lu X and Yang X 2015 Effects of atmospheric pressure cold plasma on human hepatocarcinoma cell and its 5-fluorouracil resistant cell line *Phys. Plasmas* **22** 122006

[17] Mirpour S, Ghomi H, Piroozmand S and Nikkhah M 2013 The selective characterization of nonthermal atmospheric pressure plasma jet on treatment of human breast cancer and normal cells *IEEE Trans. Plasma Sci.* **42** 315–22

[18] Torii K *et al* 2015 Effectiveness of plasma treatment on gastric cancer cells *Gastric Cancer* **18** 635–43

[19] Ja Kim S, Min Joh H and Chung T H 2013 Production of intracellular reactive oxygen species and change of cell viability induced by atmospheric pressure plasma in normal and cancer cells *Appl. Phys. Lett.* **103** 153705

[20] Keidar M, Walk R, Shashurin A, Srinivasan P, Sandler A, Dasgupta S, Ravi R, Guerrero-Preston R and Trink B 2011 Cold plasma selectivity and the possibility of a paradigm shift in cancer therapy *Br. J. Cancer* **105** 1295–301

[21] Iseki S, Nakamura K, Hayashi M, Tanaka H, Kondo H, Kajiyama H, Kano H, Kikkawa F and Hori M 2012 Selective killing of ovarian cancer cells through induction of apoptosis by nonequilibrium atmospheric pressure plasma *Appl. Phys. Lett.* **100** 113702

[22] Trink B, Keidar M, Canady J, Shamai Y and Tzukerman M 2015 The effect of cold atmospheric plasma treatment on cancer stem cells *Plasma Med* **5** 17–26

[23] Hirst A M, Frame F M, Arya M, Maitland N J and O'Connell D 2016 Low temperature plasmas as emerging cancer therapeutics: the state of play and thoughts for the future *Tumor Biol* **37** 7021–31

[24] Cheng X, Murphy W, Recek N, Yan D, Cvelbar U, Vesel A, Mozetič M, Canady J, Keidar M and Sherman J H 2014 Synergistic effect of gold nanoparticles and cold plasma on glioblastoma cancer therapy *J. Phys. D: Appl. Phys.* **47** 335402

[25] Zhu W, Lee S-J, Castro N J, Yan D, Keidar M and Zhang L G 2016 Synergistic effect of cold atmospheric plasma and drug loaded core-shell nanoparticles on inhibiting breast cancer cell growth *Sci. Rep.* **6** 21974

[26] Jalili A, Irani S and Mirfakhraie R 2016 Combination of cold atmospheric plasma and iron nanoparticles in breast cancer: gene expression and apoptosis study *Onco. Targets Ther* **9** 5911–7

[27] Kim G C, Kim G J, Park S R, Jeon S M, Seo H J, Iza F and Lee J K 2008 Air plasma coupled with antibody-conjugated nanoparticles: a new weapon against cancer *J. Phys. D: Appl. Phys.* **42** 032005

[28] He Z *et al* 2018 Cold atmospheric plasma induces ATP-dependent endocytosis of nanoparticles and synergistic U373MG cancer cell death *Sci. Rep.* **8** 5298

[29] Kumar K N, Kaushik N, Wahab R, Bhartiya P, Linh N N, Khan F, AI-Khedhairy A A and Choi E H 2020 Cold atmospheric plasma and gold quantum dots exert dual cytotoxicity mediated by the cell receptor-activated apoptotic pathway in glioblastoma cells *Cancers* **12** 457

[30] Cheng X, Rajjoub K, Shashurin A, Yan D, Sherman J H, Bian K, Murad F and Keidar M 2017 Enhancing cold atmospheric plasma treatment of cancer cells by static magnetic field *Bioelectromagnetics* **38** 53–62

[31] Wol C M, Kolb J F, Weltmann K, Woedtke T V and Bekeschus S 2020 Combination treatment with cold physical plasma and pulsed electric fields augments ROS production and cytotoxicity in lymphoma *Cancers* **12** 845

[32] Sasaki S 2018 Direct plasma stimuli including electrostimulation and OH radical induce transient increase in intracellular Ca^{2+} and uptake of a middle-size membrane-impermeable molecule *Plasma Processes Polym.* **15** 1700077

[33] Kaushik N K, Kaushik N, Yoo K C, Uddin N, Kim J S, Lee S J and Choi E H 2016 Low doses of PEG-coated gold nanoparticles sensitize solid tumors to cold plasma by blocking the PI3K/AKT-driven signaling axis to suppress cellular transformation by inhibiting growth and EMT *Biomaterials* **87** 118–30

[34] Hoon Park J *et al* 2015 A comparative study for the inactivation of multidrug resistance bacteria using dielectric barrier discharge and nano-second pulsed plasma *Sci. Rep.* **5** 13849

[35] Zhang X, Li M, Zhou R, Feng K and Yang S 2008 Ablation of liver cancer cells *in vitro* by a plasma needle *Appl. Phys. Lett.* **93** 2006–9

[36] Yan D, Sherman J H and Keidar M 2017 The application of the cold atmospheric plasma-activated solutions in cancer treatment *Anticancer Agents Med. Chem.* **18** 769–75

[37] Zhou R, Zhou R, Wang P, Xian Y and Mai-prochnow A 2020 Plasma-activated water: generation, origin of reactive species and biological *J. Phys. D: Appl. Phys.* **53** 303001

[38] Yan D, Sherman J H and Keidar M 2017 Cold atmospheric plasma, a novel promising anti-cancer treatment modality *Oncotarget* **8** 15977

[39] Kalghatgi S, Kelly C M, Cerchar E, Torabi B, Alekseev O, Fridman A, Friedman G and Azizkhan-Clifford J 2011 Effects of non-thermal plasma on mammalian cells *PLoS One* **6** e16270

[40] Kim S J, Chung T H, Bae S H and Leem S H 2010 Induction of apoptosis in human breast cancer cells by a pulsed atmospheric pressure plasma jet *Appl. Phys. Lett.* **97** 023702

[41] Thiyagarajan M, Anderson H and Gonzales X F 2014 Induction of apoptosis in human myeloid leukemia cells by remote exposure of resistive barrier cold plasma *Biotechnol. Bioeng.* **111** 565–74

[42] Han X, Klas M, Liu Y, Sharon Stack M and Ptasinska S 2013 DNA damage in oral cancer cells induced by nitrogen atmospheric pressure plasma jets *Appl. Phys. Lett.* **102** 233703

[43] Kim C H, Kwon S, Bahn J H, Lee K, Jun S I, Rack P D and Baek S J 2010 Effects of atmospheric nonthermal plasma on invasion of colorectal cancer cells *Appl. Phys. Lett.* **96** 2008–11

[44] Yan D, Talbot A, Nourmohammadi N, Cheng X, Canady J, Sherman J and Keidar M 2015 Principles of using cold atmospheric plasma stimulated media for cancer treatment *Sci. Rep.* **5** 18339

[45] Gorbanev Y, O'Connell D and Chechik V 2016 Non-thermal plasma in contact with water: the origin of species *Chem. Eur.* J **8** 15977–95

[46] Taiki I, Uchida G, Nakajima A, Takenaka K and Setsuhara Y 2016 Control of reactive oxygen and nitrogen species production in liquid by nonthermal plasma jet with controlled surrounding gas in water *Jpn. J. Appl. Phys.* **56** 01AC06

[47] Locke B R and Shih K-Y 2011 Review of the methods to form hydrogen peroxide in electrical discharge plasma with liquid water *Plasma Sources Sci. Technol.* **20** 034006

[48] Bruggeman P J *et al* 2016 Plasma–liquid interactions: a review and roadmap *Plasma Sources Sci. Technol.* **25** 053002

[49] Panngom K, Baik K Y, Nam M K, Han J H, Rhim H and Choi E H 2013 Preferential killing of human lung cancer cell lines with mitochondrial dysfunction by nonthermal dielectric barrier discharge plasma *Cell Death Dis* **4** e642

[50] Kaushik N, Kumar N, Kim C H, Kaushik N K and Choi E H 2014 Dielectric barrier discharge plasma efficiently delivers an apoptotic response in human monocytic lymphoma *Plasma Processes Polym.* **11** 1175–87

[51] Liu X, Gan L, Ma M, Zhang S and Liu J 2018 A comparative study on the transdermal penetration effect of gaseous and aqueous plasma reactive species *J. Phys. D: Appl. Phys.* **51** 075401

[52] Perinban S, Orsat V and Raghavan V 2019 Nonthermal plasma–liquid interactions in food processing: a review *Compr. Rev. Food Sci. Food Saf.* **18** 1985–2008

[53] Tanaka H, Mizuno M, Ishikawa K, Nakamura K, Kajiyama H, Kano H, Kikkawa F and Hori M 2011 Plasma-Activated medium selectively kills glioblastoma brain tumor cells by down-regulating a survival signaling molecule, AKT kinase *Plasma Med.* **1** 265–77

[54] Utsumi F, Kajiyama H, Nakamura K, Tanaka H, Mizuno M, Ishikawa K, Kondo H, Kano H, Hori M and Kikkawa F 2013 Effect of indirect nonequilibrium atmospheric pressure plasma on anti-proliferative activity against chronic chemo-resistant ovarian cancer cells *in vitro* and *in vivo PLoS One* **8** e81576

[55] Adachi T, Nonomura S, Horiba M, Hirayama T, Kamiya T, Nagasawa H and Hara H 2016 Iron stimulates plasma-activated medium-induced A549 cell injury *Sci. Rep.* **6** 20928

[56] Kumar N, Park J H, Jeon S N, Park B S, Choi E H and Attri P 2016 The action of microsecond-pulsed plasma-activated media on the inactivation of human lung cancer cells *J. Phys. D: Appl. Phys.* **49** 115401

[57] Adachi T, Tanaka H, Nonomura S, Hara H, Kondo S I and Hori M 2015 Plasma-activated medium induces A549 cell injury via a spiral apoptotic cascade involving the mitochondrial-nuclear network *Free Radic. Biol. Med.* **79** 28–44

[58] Malyavko A, Yan D, Wang Q, Klein A L, Patel K C, Sherman J H and Keidar M 2020 Cold atmospheric plasma cancer treatment, direct: versus indirect approaches *Mater. Adv.* **1** 1494–505

[59] Keidar M 2015 Plasma for cancer treatment *Plasma Sources Sci. Technol.* **24** 033001
[60] Lin L *et al* 2023 Cell membrane oscillations under radiofrequency electromagnetic modulation *Langmuir* **39** 3320 3331
[61] Yan D, Talbot A, Nourmohammadi N, Sherman J H, Cheng X and Keidar M 2015 Toward understanding the selective anticancer capacity of cold atmospheric plasma—a model based on aquaporins *Biointerphases* **10** 040801
[62] Laroussi M, Lu X and Keidar M 2017 Perspective: the physics, diagnostics, and applications of atmospheric pressure low temperature plasma sources used in plasma medicine *J. Appl. Phys.* **122** 020901
[63] Limanowski R, Yan D, Li L and Keidar M 2022 Preclinical cold atmospheric plasma cancer treatment *Cancers (Basel)* **2022** 3461
[64] Georgescu N and Lupu A R 2010 Tumoral and normal cells treatment with high-voltage pulsed cold atmosphericplasma jets *IEEE Trans. Plasma Sci.* **38** 1949–55
[65] Kim C-H, Bahn J H, Lee S-H, Kim G-Y, Jun S-I, Lee K and Baek S J 2010 Induction of cell growth arrest by atmospheric non-thermal plasma in colorectal cancer cells *J. Biotechnol.* **150** 530–8
[66] Kim G J, Kim W, Kim K T and Lee J K 2010 DNA damage and mitochondria dysfunction in cell apoptosis induced by nonthermal air plasma *Appl. Phys. Lett.* **96** 021502
[67] Ahn H J, Kim K, il, Kim G, Moon E, Yang S S and Lee J S 2011 Atmospheric-pressure plasma jet induces apoptosis involving mitochondria via generation of free radicals *PLoS One* **6** e28154
[68] Arndt S, Wacker E, Li Y F, Shimizu T, Thomas H M, Morfill G E, Karrer S, Zimmermann J L and Bosserhoff A K 2013 Cold atmospheric plasma, a new strategy to induce senescence in melanoma cells *Exp. Dermatol.* **22** 284–9
[69] Lin A *et al* 2015 Uniform nanosecond pulsed dielectric barrier discharge plasma enhances anti-tumor effects by induction of immunogenic cell death in tumors and stimulation of macrophages *Plasma Processes Polym.* **12** 1392–9
[70] Yan D, Cui H, Zhu W, Talbot A, Zhang L G, Sherman J H and Keidar M 2017 The strong cell-based hydrogen peroxide generation triggered by cold atmospheric plasma *Sci. Rep.* **7** 10831
[71] Shi L *et al* 2017 Non-thermal plasma induces a stress response in mesothelioma cells resulting in increased endocytosis, lysosome biogenesis and autophagy *Free Radic. Biol. Med.* **108** 904–17
[72] Yan D, Xu W, Yao X, Lin L, Sherman J H and Keidar M 2018 Cold atmospheric plasma cancer treatment, a critical review *Appl. Sci.* **11** 7757
[73] Yan D, Wang Q, Adhikari M, Malyavko A, Lin L, Zolotukhin D B, Yao X, Kirschner M, Sherman J H and Keidar M 2020 A physically triggered cell death via transbarrier cold atmospheric plasma cancer treatment *ACS Appl. Mater. Interfaces* **12** 34548–63
[74] Yang X *et al* 2020 Cold atmospheric plasma induces GSDME-dependent pyroptotic signaling pathway via ROS generation in tumor cells *Cell Death Dis.* **11** 295
[75] Yao X, Lin L, Soni V, Gjika E, Sherman J H, Yan D and Keidar M 2020 Sensitization of glioblastoma cells to temozolomide by a helium gas discharge tube *Phys. Plasmas* **27** 114502
[76] Keidar M, Yan D and Sherman J 2019 *Cold Plasma Cancer Therapy* (Morgan & Claypool Publishers)

[77] Gjika E, Pal-ghosh S, Tang A, Kirschner M, Tadvalkar G, Canady J, Stepp M A, Keidar M and States U 2018 Adaptation of operational parameters of cold atmospheric plasma for *in vitro* treatment of cancer cells *ACS Appl. Mater. Interfaces* **10** 9269–79

[78] Takehiko S, Yokoyama M and Johkura K 2011 A key inactivation factor of HeLa cell viability by a plasma flow *J. Phys. D: Appl. Phys.* **44** 372001

[79] Yan X, Xiong Z, Zou F, Zhao S, Lu X, Yang G, He G and Ostrikov K 2012 Plasma-induced death of HepG2 cancer cells: intracellular effects of reactive species *Plasma Processes Polym.* **9** 59–66

[80] Arjunan K P, Friedman G, Fridman A and Clyne A M 2012 Non-thermal dielectric barrier discharge plasma induces angiogenesis through reactive oxygen species *J. R. Soc. Interface* **9** 147–57

[81] Ahn H J, Kim K I, Hoan N N, Kim C H, Moon E, Choi K S, Yang S S and Lee J S 2014 Targeting cancer cells with reactive oxygen and nitrogen species generated by atmospheric-pressure air plasma *PLoS One* **9** e86173

[82] Kim S J and Chung T H 2016 Cold atmospheric plasma jet-generated RONS and their selective effects on normal and carcinoma cells *Sci. Rep.* **6** 20332

[83] Conway G E, Casey A, Milosavljevic V, Liu Y, Howe O, Cullen P J and Curtin J F 2016 Non-thermal atmospheric plasma induces ROS-independent cell death in U373MG glioma cells and augments the cytotoxicity of temozolomide *Br. J. Cancer* **114** 435–43

[84] Nguyen N H, Park H J, Yang S S, Choi K S and Lee J-S 2016 Anti-cancer efficacy of nonthermal plasma dissolved in a liquid, liquid plasma in heterogeneous cancer cells *Sci. Rep.* **6** 29020

[85] Liu Y, Tan S, Zhang H, Kong X, Ding L, Shen J and Lan Y 2017 Selective effects of non-thermal atmospheric plasma on triple-negative breast normal and carcinoma cells through different cell signaling pathways *Sci. Rep.* **7** 7908

[86] Bauer G and Graves D B 2016 Mechanisms of selective antitumor action of cold atmospheric plasma-derived reactive oxygen and nitrogen species *Plasma Processes Polym.* **13** 1157–78

[87] Neha K, Uddin N, Sim G B, Hong Y J, Baik K Y, Kim C H, Lee S J, Kaushik N K and Choi E H 2015 Responses of solid tumor cells in DMEM to reactive oxygen species generated by non-thermal plasma and chemically induced ROS systems *Sci. Rep.* **5** 8587

[88] Ishaq M, Bazaka K and Ostrikov K 2015 Pro-apoptotic NOXA is implicated in atmospheric-pressure plasma-induced melanoma cell death *J. Phys. D: Appl. Phys.* **48** 464002

[89] Adhikari M, Adhikari B, Ghimire B, Baboota S and Choi E H 2020 Cold atmospheric plasma and silymarin nanoemulsion activate autophagy in human melanoma cells *Int. J. Mol. Sci.* **21** 1939

[90] Saito K, Asai T, Fujiwara K, Sahara J, Koguchi H, Fukuda N, Suzuki-Karasaki M, Soma M and Suzuki-Karasaki Y 2016 Tumor-selective mitochondrial network collapse induced by atmospheric gas plasma-activated medium *Oncotarget* **7** 19910

[91] Bekeschus S, Wende K, Hefny M M, R?dder K, Jablonowski H, Schmidt A, Woedtke T von, Weltmann K-D and Benedikt J 2017 Oxygen atoms are critical in rendering THP-1 leukaemia cells susceptible to cold physical plasma-induced apoptosis *Sci. Rep.* **7** 2791

[92] Li W, Yu K N, Ma J, Shen J, Cheng C, Zhou F, Cai Z and Han W 2017 Non-thermal plasma induces mitochondria-mediated apoptotic signaling pathway via ROS generation in HeLa cells *Arch. Biochem. Biophys.* **633** 68–77

[93] Lin A, Truong B, Patel S, Kaushik N, Choi E H, Fridman G, Fridman A and Miller V 2017 Nanosecond-pulsed dbd plasma-generated reactive oxygen species trigger immunogenic cell death in a549 lung carcinoma cells through intracellular oxidative stress *Int. J. Mol. Sci.* **18** 966

[94] Ishaq M, Evans M D M and Ostrikov K K 2014 Atmospheric pressure gas plasma-induced colorectal cancer cell death is mediated by Nox2-ASK1 apoptosis pathways and oxidative stress is mitigated by Srx-Nrf2 anti-oxidant system *Biochim. Biophys. Acta, Mol. Cell. Res.* **1843** 2827–37

[95] Ishaq M, Kumar S, Varinli H, Han Z J, Rider A E, Evans M D M, Murphy A B and Ostrikov K 2014 Atmospheric gas plasma-induced ROS production activates TNF-ASK1 pathway for the induction of melanoma cancer cell apoptosis *Mol. Biol. Cell* **25** 1523–31

[96] Bekeschus S, Kolata J, Winterbourn C, Kramer A, Turner R, Weltmann K D, Br B and Masur K 2014 Hydrogen peroxide: a central player in physical plasma-induced oxidative stress in human blood cells *Free Radic. Res.* **48** 542–9

[97] Ninomiya K, Ishijima T, Imamura M, Yamahara T, Enomoto H, Takahashi K, Tanaka Y, Uesugi Y and Shimizu N 2013 Evaluation of extra- and intracellular OH radical generation, cancer cell injury, and apoptosis induced by a non-thermal atmospheric-pressure plasma jet *J. Phys. D: Appl. Phys.* **46** 425401

[98] Moncada S and Erusalimsky J D 2002 Does nitric oxide modulate mitochondrial energy generation and apoptosis? *Nat. Rev. Mol. Cell Biol.* **3** 214–20

[99] Li Y, Ho Kang M, Sup Uhm H, Joon Lee G, Ha Choi E and Han I 2017 Effects of atmospheric-pressure non-thermal bio-compatible plasma and plasma activated nitric oxide water on cervical cancer cells *Sci. Rep.* **7** 45781

[100] Galván A and Fernández E 2001 Eukaryotic nitrate and nitrite transporters *Cell. Mol. Life Sci.* **58** 225–33

[101] Ikeda M, Beitz E, Kozono D, Guggino W B, Agre P and Yasui M 2002 Characterization of aquaporin-6 as a nitrate channel in mammalian cells. Requirement of pore-lining residue threonine 63 *J. Biol. Chem.* **277** 39873–9

[102] Marla S S, Lee J and Groves J T 1997 Peroxynitrite rapidly permeates phospholipid membranes *Proc. Natl Acad. Sci. USA* **94** 14243–8

[103] Denicola A, Souza J M and Radi R 1998 Diffusion of peroxynitrite across erythrocyte membranes *Proc. Natl Acad. Sci. USA* **95** 3566–71

[104] Agre P 2004 Aquaporin water channels (nobel lecture) *Angew. Chem. Int. Ed.* **43** 4278–90

[105] Wu B and Beitz E 2007 Aquaporins with selectivity for unconventional permeants *Cell. Mol. Life Sci.* **64** 2413–21

[106] Bienert G P and Chaumont F 2014 Aquaporin-facilitated transmembrane diffusion of hydrogen peroxide *Biochim. Biophys. Acta, Gen. Subj.* **1840** 1596–604

[107] Miller E W, Dickinson B C and Chang C J 2010 Aquaporin-3 mediates hydrogen peroxide uptake to regulate downstream intracellular signaling *Proc. Natl Acad. Sci. USA* **107** 15681–6

[108] Almasalmeh A, Krenc D, Wu B and Beitz E 2014 Structural determinants of the hydrogen peroxide permeability of aquaporins *FEBS J.* **281** 647–56

[109] Kawasaki T, Kusumegi S, Kudo A, Sakanoshita T, Tsurumaru T, Sato A, Uchida G, Koga K and Shiratani M 2016 Effects of irradiation distance on supply of reactive oxygen species to the bottom of a Petri dish filled with liquid by an atmospheric O_2/He plasma jet *J. Appl. Phys.* **119** 173301

[110] Herrera M, Hong N J and Garvin J L 2006 Aquaporin-1 transports NO across cell membranes *Hypertension* **48** 157–64

[111] Yan D, Xiao H, Zhu W, Nourmohammadi N, Zhang L G, Bian K and Keidar M 2017 The role of aquaporins in the anti-glioblastoma capacity of the cold plasma-stimulated medium *J. Phys. D: Appl. Phys.* **50** 055401

[112] Yusupov M, Yan D, Cordeiro R M and Bogaerts A 2018 Atomic scale simulation of H_2O_2 permeation through aquaporin: toward the understanding of plasma cancer treatment *J. Phys. D: Appl. Phys.* **51** 125401

[113] Cox A G, Winterbourn C C and Hampton M B 2010 Mitochondrial peroxiredoxin involvement in antioxidant defence and redox signalling *Biochem.* J **425** 313–25

[114] Sandalova T, Zhong L, Lindqvist Y, Holmgren A and Schneider G 2001 Three-dimensional structure of a mammalian thioredoxin reductase: implications for mechanism and evolution of a selenocysteine-dependent enzyme *Proc. Natl Acad. Sci. USA* **98** 9533–8

[115] Fritz-Wolf K, Kehr S, Stumpf M, Rahlfs S and Becker K 2011 Crystal structure of the human thioredoxin reductase–thioredoxin complex *Nat. Commun.* **2** 383

[116] Andreyev A Y, Kushnareva Y E and Starkov A 2005 Mitochondrial metabolism of reactive oxygen species *Biochemistry (Mosc)* **70** 200–14

[117] Kaushik N K, Kaushik N, Park D and Choi E H 2014 Altered antioxidant system stimulates dielectric barrier discharge plasma-induced cell death for solid tumor cell treatment *PLoS One* **9** e103349

[118] Weiss M *et al* 2015 Cold atmospheric plasma treatment induces anti-proliferative effects in prostate cancer cells by redox and apoptotic signaling pathways *PLoS One* **10** e0130350

[119] Nimse S B and Pal D 2015 Free radicals, natural antioxidants, and their reaction mechanism *RSC Adv.* **5** 27986–8006

[120] Finkel T 2011 Signal transduction by reactive oxygen species *J. Cell Biol.* **194** 7–15

[121] Article E, Van Der Paal J, Neyts E C, Verlackt C C W and Bogaerts A 2015 Permeability of cancer and normal cells subjected to oxidative stress *Chem. Sci.* **7** 489–98

[122] Razzokov J, Yusupov M, Cordeiro R M and Bogaerts A 2018 Atomic scale understanding of the permeation of plasma species across native and oxidized membranes *J. Phys. D: Appl. Phys.* **51** 365203

[123] Oliveira M C, Yusupov M, Bogaerts A and Cordeiro R M 2020 How do nitrated lipids affect the properties of phospholipid membranes? *Arch. Biochem. Biophys.* **695** 108548

[124] Van Der Paal J, Neyts E C, Verlackt C C W and Bogaerts A 2016 Effect of lipid peroxidation on membrane permeability of cancer and normal cells subjected to oxidative stress *Chem. Sci.* **7** 489–98

[125] Ray A, Ranieri P, Karamchand L, Yee B, Foster J and Kopelman R 2017 Real-time monitoring of intracellular chemical changes in response to plasma irradiation *Plasma Med* **7** 7–26

[126] Chang J W, Kang S U, Shin Y S, Kim K I I, Seo S J, Yang S S, Lee J S, Moon E, Lee K and Kim C H 2014 Non-thermal atmospheric pressure plasma inhibits thyroid papillary cancer cell invasion via cytoskeletal modulation, altered MMP-2/-9/uPA activity *PLoS One* **9** e92198

[127] Recek N, Cheng X, Keidar M, Cvelbar U, Vesel A, Mozetic M and Sherman J 2015 Effect of cold plasma on glial cell morphology studied by atomic force microscopy *PLoS One* **10** e00119111

[128] Shashurin A, Stepp M A, Hawley T S, Pal-Ghosh S, Brieda L, Bronnikov S, Jurjus R A and Keidar M 2010 Influence of cold plasma atmospheric jet on surface integrin expression of living cells *Plasma Processes Polym.* **7** 294–300

[129] Lee J, Om J, Kim Y, Kim K and Choi E 2016 Selective killing effects of cold atmospheric pressure plasma with NO induced dysfunction of epidermal growth factor receptor in oral squamous cell carcinoma *PLoS One* **11** e0150279
[130] O'connell D, Cox L J, Hyland W B, McMachon S J, Reuter S, Grahma W G, Gans T and Currell F J 2015 Cold atmospheric pressure plasma jet interactions with plasmid DNA *Appl. Phys. Lett.* **98** 043701
[131] Kastan M B 2008 DNA damage responses: mechanisms and roles in human disease: 2007 G.H.A. Clowes Memorial Award Lecture *Mol. Cancer Res.* **6** 517–24
[132] Chang J W *et al* 2014 Non-thermal atmospheric pressure plasma induces apoptosis in oral cavity squamous cell carcinoma: involvement of DNA-damage-triggering sub-G1 arrest via the ATM/p53 pathway *Arch. Biochem. Biophys.* **545** 133–40
[133] Ishaq M, Bazaka K and Ostrikov K 2015 Intracellular effects of atmospheric-pressure plasmas on melanoma cancer cells *Phys. Plasmas* **22** 122003
[134] Joerger A C and Fersht A R 2008 Structural biology of the tumor suppressor p53 *Annu. Rev. Biochem.* **77** 557–82
[135] Volotskova O, Hawley T S, Stepp M A and Keidar M 2012 Targeting the cancer cell cycle by cold atmospheric plasma *Sci. Rep.* **2** 636
[136] Gherardi M, Turrini E, Laurita R, De Gianni E, Ferruzzi L, Liguori A, Stancampiano A, Colombo V and Fimognari C 2015 Atmospheric non-equilibrium plasma promotes cell death and cell-cycle arrest in a lymphoma cell line *Plasma Processes Polym.* **12** 1354–63
[137] Lee J-H, Om J-Y, Kim Y-H, Kim K-M, Choi E-H and Kim K-N 2016 Selective killing effects of cold atmospheric pressure plasma with NO induced dysfunction of epidermal growth factor receptor in oral squamous cell carcinoma *PLoS One* **11** e0150279
[138] Ratovitski E A, Cheng X, Yan D, Sherman J H, Canady J, Trink B and Keidar M 2014 Anti-cancer therapies of 21st century: novel approach to treat human cancers using cold atmospheric plasma *Plasma Processes Polym.* **11** 1128–37
[139] Partecke L I *et al* 2012 Tissue tolerable plasma (TTP) induces apoptosis in pancreatic cancer cells *in vitro* and *in vivo BMC Cancer* **12** 473
[140] Naciri M, Dowling D and Al-Rubeai M 2014 Differential sensitivity of mammalian cell lines to non-thermal atmospheric plasma *Plasma Processes Polym.* **11** 391–400
[141] Kim J Y, Kim S O, Wei Y and Li J 2010 A flexible cold microplasma jet using biocompatible dielectric tubes for cancer therapy *Appl. Phys. Lett.* **96** 203701
[142] Zucker S N *et al* 2012 Preferential induction of apoptotic cell death in melanoma cells as compared with normal keratinocytes using a non-thermal plasma torch *Cancer Biol. Ther.* **4047** 1299–306
[143] Yan N and Shi Y 2005 Mechanisms of apoptosis through structural biology *Annu. Rev. Cell Dev. Biol.* **21** 35–56
[144] Mattson M P 2000 Apoptosis in neurodegenerative disorders *Nat. Rev. Mol. Cell Biol.* **1** 120–9
[145] Jiang X and Wang X 2004 Cytochrome *C*-mediated apoptosis *Annu. Rev. Biochem.* **73** 87–106
[146] Elmore S 2007 A review of programmed cell death *Toxicol. Pathol.* **35** 495–516
[147] Riedl S J and Salvesen G S 2007 The apoptosome: signalling platform of cell death *Nat. Rev. Mol. Cell Biol.* **8** 405–13
[148] Jacobson M D, Weil M and Raff M C 1997 Programmed cell death in animal development *Cell* **88** 347–54

[149] Kang S U *et al* 2014 Nonthermal plasma induces head and neck cancer cell death: the potential involvement of mitogen-activated protein kinase-dependent mitochondrial reactive oxygen species *Cell Death Dis* **5** e1056
[150] Ma Y, Ha C S, Hwang S W, Lee H J, Kim G C, Lee K W and Song K 2014 Non-thermal atmospheric pressure plasma preferentially induces apoptosis in p53-mutated cancer cells by activating ROS stress-response pathways *PLoS One* **9** e91947
[151] Gibson A R, McCarthy H O, Ali A A, O'Connell D and Graham W G 2014 Interactions of a non-thermal atmospheric pressure plasma effluent with PC-3 prostate cancer cells *Plasma Processes Polym.* **11** 1142–9
[152] Neha K, Lee S J, Choi T G, Baik K Y, Uhm H S, Kim C H, Kaushik N K and Choi E H 2015 Non-thermal plasma with 2-deoxy-D-glucose synergistically induces cell death by targeting glycolysis in blood cancer cells *Sci. Rep.* **5** 8726
[153] Kim S J and Chung T H 2015 Plasma effects on the generation of reactive oxygen and nitrogen species in cancer cells in-vitro exposed by atmospheric pressure pulsed plasma jets *Appl. Phys. Lett.* **107** 063702
[154] Kaushik N, Lee S J, Choi T G, Baik K Y, Uhm H S, Kim C H, Kaushik N K and Choi E H 2021 Glycolytic inhibitor induces metabolic crisis in solid cancer cells to enhance cold plasma-induced cell death *Plasma Processes Polym.* **18** 2000187
[155] Schneider C, Gebhardt L, Arndt S, Karrer S, Zimmermann J L, Fischer M J M and Bosserhoff A-K 2018 Cold atmospheric plasma causes a calcium influx in melanoma cells triggering CAP-induced senescence *Sci. Rep.* **8** 10048
[156] Wang C, Zhang H, Xue Z, Yin H, Niu Q and Chen H 2015 The relation between doses or post-plasma time points and apoptosis of leukemia cells induced by dielectric barrier discharge plasma *AIP Adv.* **5** 127220
[157] Hotchkiss R S 2009 Cell death *N. Engl. J. Med.* **361** 1570–83
[158] Hansen M, Rubinsztein D C and Walker D W 2018 Autophagy as a promoter of longevity: insights from model organisms *Nat. Rev. Mol. Cell Biol.* **19** 579–93
[159] Mariño G, Niso-Santano M, Baehrecke E H and Kroemer G 2014 Self-consumption: the interplay of autophagy and apoptosis *Nat. Rev. Mol. Cell Biol.* **15** 81–94
[160] Debnath J, Baehrecke E H and Kroemer G 2005 Does autophagy contribute to cell death? *Autophagy* **1** 66–74
[161] Gozuacik D and Kimchi A 2004 Autophagy as a cell death and tumor suppressor mechanism *Oncogene* **23** 2891–906
[162] Lin L and Baehrecke E H 2015 Autophagy, cell death, and cancer *Mol. Cell Oncol.* **2** e985913
[163] Li L, Ishdorj G and Gibson S B 2012 Reactive oxygen species regulation of autophagy in cancer: implications for cancer treatment *Free Radic. Biol. Med.* **53** 1399–410
[164] Yoshikawa N *et al* 2020 Plasma-activated medium promotes autophagic cell death along with alteration of the mTOR pathway *Sci. Rep.* **10** 1614
[165] Lukes P, Dolezalova E, Sisrova I and Clupek M 2014 Aqueous-phase chemistry and bactericidal effects from an air discharge plasma in contact with water: evidence for the formation of peroxynitrite through a pseudo-second-order post-discharge reaction of H_2O_2 and HNO_2 *Plasma Sources Sci. Technol.* **23** 015019
[166] Le Bras M, Clément M V, Pervaiz S and Brenner C 2005 Reactive oxygen species and the mitochondrial signaling pathway of cell death *Histol. Histopathol.* **20** 205–20

[167] Padmaja S, Squadrito G L and Pryor W A 1998 Inactivation of glutathione peroxidase by peroxynitrite *Arch. Biochem. Biophys.* **349** 1–6
[168] Peshenko I V and Shichi H 2001 Oxidation of active center cysteine of bovine 1-Cys peroxiredoxin to the cysteine sulfenic acid form by peroxide and peroxynitrite *Free Radic. Biol. Med.* **31** 292–303
[169] López-Lázaro M 2007 Dual role of hydrogen peroxide in cancer: possible relevance to cancer chemoprevention and therapy *Cancer Lett.* **252** 1–8
[170] Henle E S and Linn S 1997 Formation, prevention, and repair of DNA damage by iron/hydrogen peroxide *J. Biol. Chem.* **272** 19095–8
[171] Skovsen E, Snyder J W, Lambert J D C and Ogilby P R 2005 Lifetime and diffusion of singlet oxygen in a cell *J. Phys. Chem.* B **109** 8570–3
[172] Lindig B A, Rodgers M A J and Schaaplc A P 1980 Determination of the lifetime of singlet oxygen in D20 using 9,10-anthracenedipropionic acid, a water-soluble probe *J. Am. Chem. Soc.* **102** 5590–3
[173] Fransen M, Nordgren M, Wang B and Apanasets O 2012 Role of peroxisomes in ROS/RNS-metabolism: implications for human disease *Biochim. Biophys. Acta, Mol. Basis Dis.* **1822** 1363–73
[174] Yan D, Horkowitz A, Wang Q and Keidar M 2021 On the selective killing of cold atmospheric plasma cancer treatment: status and beyond *Plasma Processes Polym.* **18** 2100020
[175] Pryor W A 1986 Oxy-radicals and related and reactions *Annu. Rev. Physiol.* **48** 657–67
[176] Janik I and Tripathi G N R 2013 The nature of the superoxide radical anion in water *J. Chem. Phys.* **139** 014302
[177] Yan D, Nourmohammadi N, Bian K, Murad F, Sherman J H and Keidar M 2016 Stabilizing the cold plasma-stimulated medium by regulating medium's composition *Sci. Rep.* **6** 26016
[178] Koppenol H and Koppenol H 1996 Nitric oxide, superoxide, and peroxynitrite: the good, the bad, and the ugly *Am. J. Physiol. Cell Physiol.* **271** 1424–37
[179] Breen C, Pal R, Elsegood M R J, Teat S J, Iza F, Wende K, Buckley B R and Butler S J 2020 Time-resolved luminescence detection of peroxynitrite using a reactivity-based lanthanide probe *Chem. Sci.* **11** 3164–70
[180] Yan D, Nourmohammadi N, Milberg J and Sherman J H 2018 Guidelines for using 3-nitro-L-tyrosine as an antidegradation reagent of H_2O_2 in the cold atmospheric plasma-stimulated solutions *Plasma Med* **8** 121–30
[181] Yan D, Sherman J H, Cheng X, Ratovitski E, Canady J and Keidar M 2014 Controlling plasma stimulated media in cancer treatment application *Appl. Phys. Lett.* **105** 224101
[182] Takai E, Kitamura T, Kuwabara J, Ikawa S, Yoshizawa S, Shiraki K, Kawasaki H, Arakawa R and Kitano K 2014 Chemical modification of amino acids by atmospheric-pressure cold plasma in aqueous solution *J. Phys. D: Appl. Phys.* **47** 285403
[183] Yokoyama M, Johkura K and Sato T 2014 Gene expression responses of HeLa cells to chemical species generated by an atmospheric plasma flow *Biochem. Biophys. Res. Commun.* **450** 1266–71
[184] Mohades S, Laroussi M, Sears J, Barekzi N and Razavi H 2015 Evaluation of the effects of a plasma activated medium on cancer cells *Phys. Plasmas* **22** 122001

[185] Judée F, Fongia C, Ducommun B, Yousfi M, Lobjois V and Merbahi N 2016 Short and long time effects of low temperature plasma activated media on 3D multicellular tumor spheroids *Sci. Rep.* **6** 21421
[186] Bauder G 2016 The antitumor effect of singlet oxygen *Anticancer Res.* **36** 5649–64
[187] Liu Z, Xu D, Liu D, Cui Q and Cai H 2017 Production of simplex RNS and ROS by nanosecond pulse N_2/O_2 plasma jets with homogeneous shielding gas for inducing myeloma cell apoptosis *J. Phys. D: Appl. Phys.* **50** 195204
[188] Yan D, Cui H, Zhu W, Nourmohammadi N, Milberg J, Zhang L G, Sherman J H and Keidar M 2017 The specific vulnerabilities of cancer cells to the cold atmospheric plasma-stimulated solutions *Sci. Rep.* **7** 4479
[189] Tanaka H *et al* 2015 Plasma with high electron density and plasma-activated medium for cancer treatment *Clin. Plasma Med.* **3** 72–6
[190] Yan D, Cui H, Zhu W, Nourmohammadi N, Milberg J, Zhang L G, Sherman J H and Keidar M 2016 Temporal evaluation of the anti-tumor efficiency of plasma-activated media *Plasma Processes Polym.* **13** 1206–11
[191] Saadati F, Mahdikia H, Abbaszadeh H and Abdollahifar M 2018 Comparison of direct and indirect cold atmospheric-pressure plasma methods in the $B_{16}F_{10}$ melanoma cancer cells treatment *Sci. Rep.* **8** 7689
[192] Yan D, Xu W, Yao X, Lin L, Sherman J H and Keidar M 2018 The cell activation phenomena in the cold atmospheric plasma cancer treatment *Sci. Rep.* **8** 15418
[193] Garate J A, English N J and MacElroy J M D 2011 Human aquaporin 4 gating dynamics in dc and ac electric fields: a molecular dynamics study *J. Chem. Phys.* **134** 02B613
[194] English N J and Garate J A 2016 Near-microsecond human aquaporin 4 gating dynamics in static and alternating external electric fields: non-equilibrium molecular dynamics *J. Chem. Phys.* **145** 085102
[195] Yan D, Lin L, Xu W, Nourmohammadi N and Jonathan H 2018 Universality of micromolar-level cell-based hydrogen peroxide generation during direct cold atmospheric plasma treatment *Plasma Med.* **8** 335–43
[196] Szatrowski T P and Nathan C F 1991 Production of large amounts of hydrogen peroxide by human tumor cells *Cancer Res.* **51** 794–9
[197] Zieba M, Suwalski M, Kwiatkowska S, Piasecka G, Grzelewskarzymowska I, Stolarek R and Nowak D 2000 Comparison of hydrogen peroxide generation and the content of lipid peroxidation products in lung cancer tissue and pulmonary parenchyma *Respir. Med.* **94** 800–5
[198] Lim S D, Sun C, Lambeth J D, Marshall F, Amin M, Chung L, Petros J A and Arnold R S 2005 Increased Nox1 and hydrogen peroxide in prostate cancer *Prostate* **62** 200–7
[199] Chen Q, Espey M G, Krishna M C, Mitchell J B, Corpe C P, Buettner G R, Shacter E and Levine M 2005 Pharmacologic ascorbic acid concentrations selectively kill cancer cells: action as a pro-drug to deliver hydrogen peroxide to tissues *Proc. Natl Acad. Sci. USA* **102** 13604–9
[200] Nozik-Grayck E, Suliman H B and Piantadosi C A 2005 Extracellular superoxide dismutase *Int. J. Biochem. Cell Biol.* **37** 2466–71
[201] Ameziane-El-Hassani R, Schlumberger M and Dupuy C 2016 NADPH oxidases: new actors in thyroid cancer? *Nat. Rev. Endocrinol.* **12** 485–94
[202] O'Leary B R *et al* 2015 Loss of SOD3 (EcSOD) expression promotes an aggressive phenotype in human pancreatic ductal adenocarcinoma *Clin. Cancer Res.* **21** 1741–51

[203] Singh B and Bhat H K 2012 Superoxide dismutase 3 is induced by antioxidants, inhibits oxidative DNA damage and is associated with inhibition of estrogen-induced breast cancer *Carcinogenesis* **33** 2601–10
[204] Teoh M L T, Fitzgerald M P, Oberley L W and Domann F E 2009 Overexpression of extracellular superoxide dismutase attenuates heparanase expression and inhibits breast carcinoma cell growth and invasion *Cancer Res.* **69** 6355–63

Chapter 5

Cold atmospheric plasma-activated solutions

Over the past several years, cold atmospheric plasma (CAP)-activated solutions (PAS) have shown promising applications in cancer treatment. PAS also shows a selective anti-cancer capacity. Furthermore, compared with the direct treatment, PAS can be stored for a relatively long time and can be used in many cases without dependence on a CAP source. Research on PAS is a hot topic in plasma medicine. In chapter 5, we comprehensively introduce the key PAS topics.

5.1 Indirect CAP treatment

Two general strategies for using CAP in cancer treatment have been developed. One is by directly employing plasma generated from the plasma sources to treat the cells grown *in vitro* or the subcutaneous tumors grown in mice [1–3]. In most studies, cancer or normal cells are immersed in a thin layer of cell culture medium or other biologically adaptable solutions [4–6]. Such a layer of solutions facilitates the transition of the plasma-originated species from the gas phase into the aqueous phase [7]. Another method is employing the plasma-activated biologically adaptable solutions (PAS) such as medium, phosphate-buffered saline, or Ringer's solutions to trigger the death of cancer cells or inhibit tumorous tissue growth via subcutaneous or intraperitoneal injection [8]. Early studies used deionized water to make PAS, however, the clinical utility is limited. In addition, deionized water is neither suitable for cell culture *in vitro* nor for animal study *in vivo*.

Typically, PAS can be made by treating biologically adaptable solutions such as medium or phosphate-buffered saline (PBS) with a CAP source above solution [10–16] (figure 5.1). When a CAP source such as an Atmospheric Pressure Plasma Jet (APPJ) is used to activate solutions, a visible plasma plume does not even need to touch the solution surface, because the gas flow will still carry reactive species allowing them to enter solution (figure 5.2) [17, 18]. In this setting, reactive species concentration and

doi:10.1088/978-0-7503-5537-7ch5

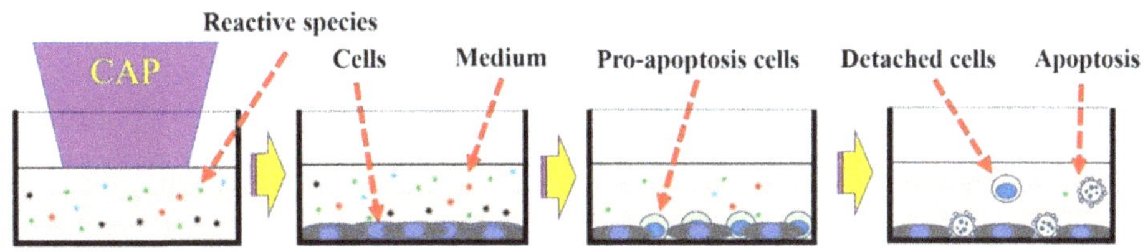

Figure 5.1. Schematic illustration of the indirect treatment *in vitro*. Here, apoptosis represents one of many cell death mechanisms. Reproduced from [9]. CC BY 4.0.

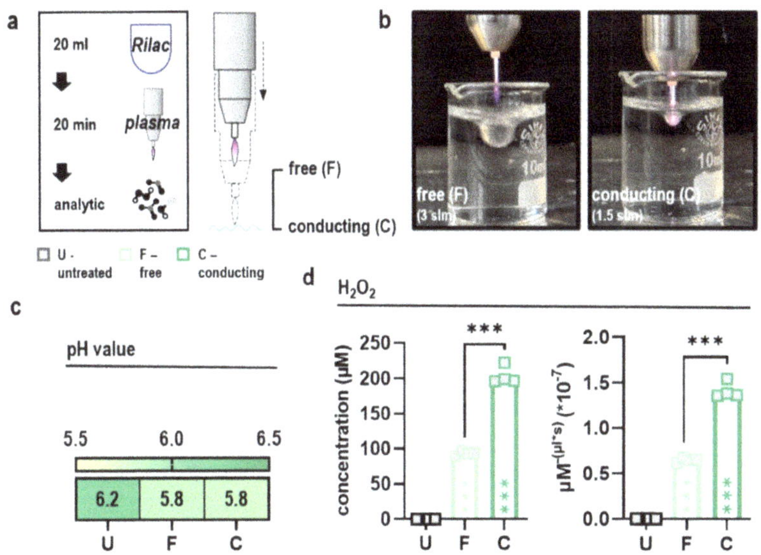

Figure 5.2. Conducting CAP treatments delivered more reactive species into Ringer's lactate solutions than the treatment with a gap between observable bulk plasma and solution surface [18]. (a) Schematic overview of CAP treatment on Ringer's lactate (Rilac) in free (F) and conducting mode (C). (b) Photos of F (3 slm) and C (1.5 slm) model of treatment. (c) pH in the CAP-activated Ringer's lactate solutions (PAL). (d) H_2O_2 concentration and delivery rate in PAL: U (control), F, and C model. The heat map in C shows medians. Bar graphs show medians and individual values. Reproduced from [18]. CC BY 4.0.

corresponding anti-cancer performance will be compromised [17, 18]. Moreover, PAS can also be made by the direct discharge in these solutions [19, 20]. Long-lived reactive species, such as H_2O_2, NO_2^-, NO_3^-, and $ONOO^-$, have been found in PAS [21, 22]. Chemical components in PAS can be effectively modulated by controlling the discharge conditions such as the carrying gas input in an APPJ source, Ar versus He/O_2, as shown in figure 5.3. The anti-cancer effect of PAS is based on the long-lived reactive species and other products formed in the reaction between plasma and the original components in solutions, such as the amino acids in the medium and the lactate in Ringer's solution [10, 12, 13, 23–28]. PAS can be stored under certain conditions for at least a week [29–31].

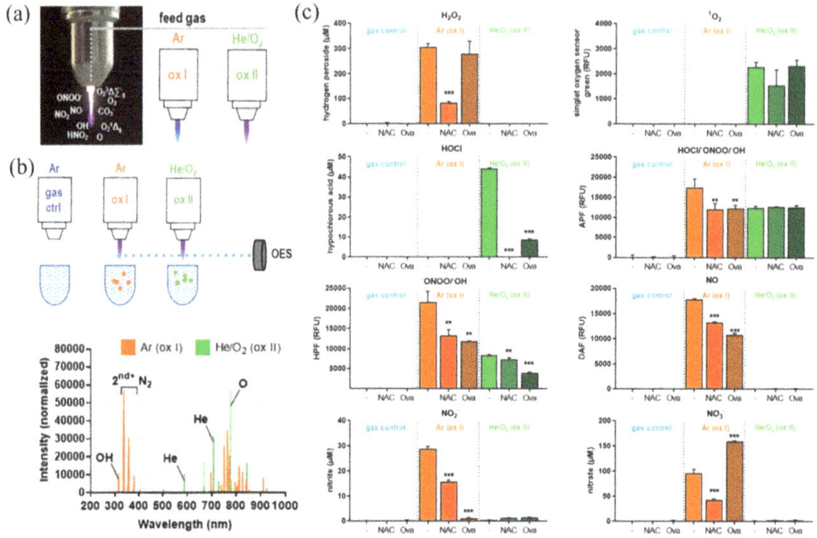

Figure 5.3. ROS/RNS fingerprint in the APPJ-treated PBS. (a) APPJ source (kINPen) with two carrying gas inputs. (b) Optical emission spectroscopy (OES) measurement of plasma plume. Peaks represent different types of atoms, molecules, such as N_2, O, and N_2. (c) Some reactive species in the treated PBS with the presence or absence of ovalbumin (Ova) or n-acetylcysteine (NAC). Ar = argon; He = helium; crtl = control. Reproduced from [32]. CC BY 4.0.

This may be a critical advantage of indirect treatment, which allows CAP to be adapted as a pharmacological method.

Long-lived reactive species, such as H_2O_2, mainly originate from the short-lived reactive species in the gas phase, such as OH. radical. Methylene blue (MB) is highly reactive with hydroxyl radicals (OH.) in aqueous solutions [15–18], but not highly reactive with H_2O_2. The effect of MB on H_2O_2 generation in the CAP-activated PBS was investigated by increasing MB's concentration in PBS. One can see that a high concentration of MB strongly inhibits H_2O_2 generation in the CAP-activated PBS. When MB concentration increases to 100 μM, H_2O_2 concentration in the CAP-treated MB-containing PBS and the H_2O_2-treated MB-containing PBS was 10.4 and 32.9 μM, respectively. That means about 22.5 μM of short-lived reactive species based H_2O_2 generation has been inhibited by 100 μM MB. Considering the H_2O_2 concentration of the CAP-activated 0 μM MB-containing PBS was about 47.4 μM, OH. contributes to at least 47.5% of total H_2O_2 generation [31].

To date, the CAP-activated medium (PAM) is the most extensively investigated PAS. The early demonstration of the cytotoxicity of PAM was reported in the study by Kalghatgi *et al* which revealed DNA double-strand breaks (DSB) caused by DBD treatment on breast epithelial cells (MCF10A) [33]. The indirect CAP treatment on medium could also generate DSB on MCF10A cells, like the direct CAP treatment on cells. This study is one of the earliest works to demonstrate the critical role of the CAP-originated species in the cytotoxicity of CAP treatment *in vitro* [33]. Subsequently, Takehiko Sato, *et al* revealed the cytotoxicity of PAM over HeLa cells [13]. They also found that H_2O_2 was one of the main factors responsible for the inactivation of HeLa cells by PAM [13]. In addition, Hiromasa Tanaka *et al* demonstrated the selective anti-cancer capacity of PAM over glioblastoma (U251SP) cells [10]. Their pioneering work revealed the basic features of PAM that other scientists repeatedly found in the years following. First, the cytotoxicity of PAM is inversely proportional to the density of cancer cells [10]. PAM made by a relatively short CAP treatment will only inhibit the growth of cancer cells with a relatively low density [10]. In addition, they discovered that the cytotoxicity of PAM was gradually lost following storage at room temperature [10]. More importantly, PAS has shown selective anti-cancer performance [10, 15, 19]. For example, PAM could selectively kill glioblastoma cells rather than normal astrocytes (ACBRI-371) [10]. After that, a series of studies validated these results (table 5.1). PAM selectively kills cancer cells from lung, colon, melanoma, cervical, breast, osteosarcoma, bladder, as well as liver cancer [14, 15, 19, 34–38].

Table 5.1. Examples of using PAS in cancer treatment *in vitro*. DBD: dielectric barrier discharge; PBS: phosphate-buffered saline; Others: Discharge in the medium; Tumor spheroid: Colon adenocarcinoma multicellular tumor spheroid; FBS: fetal bovine serum; Hamster cells: Chinese hamster cells. I just enumerated the studies using the biologically adaptable solutions here.

Years	CAP	Solutions	Cancer cell lines	Normal cell lines	Selectivity	Reference
2011	Plasma jet	Medium	Glioblastoma (U251SP)	Astrocytes (ACBRI-371)	Yes	[10]
2011	Plasma jet	Medium	Cervical cancer (HeLa)		N/A	[13]
2012	DBD	Medium	Glioblastoma (U87MG)		N/A	[11]
2012	DBD	Medium	Colon cancer (HCT116)		N/A	[11]
2013	Plasma jet	Medium	Ovarian cancer (NOS2/3)		N/A	[49]
2014	DBD	Medium	Cervical cancer (HeLa)		N/A	[28]
2014	Plasma jet	Medium	Glioblastoma (U87MG)		N/A	[12]
2015	Plasma jet	Medium	Lung cancer (A549)		N/A	[43]
2015	Plasma jet	Medium	Liver cancer (HepG2)		N/A	[43]
2015	Plasma jet	Medium	Breast cancer (MCF-7)		N/A	[43]
2015	Plasma jet	Medium	Glioblastoma (U87MG)		N/A	[30]
2015	Plasma jet	Medium	Breast cancer (MCF-7)		N/A	[30]
2015	Plasma jet	Medium	Breast cancer (MDA-MB-231)		N/A	[30]
2015	Plasma jet	Medium	Bladder cancer (SCaBER)		N/A	[44]
2015	DBD	Medium	Hepatocellular carcinoma (Bel7402)		N/A	[50]
2016	Plasma jet	Medium/PBS	Glioblastoma (U87MG)		N/A	[45]

2016	Plasma jet	Medium/PBS	Pancreatic cancer (PA-TU-8988T)		N/A	[45]
2016	Plasma jet	Medium/PBS	Breast cancer (MDA-MB-231)		N/A	[45]
2016	Plasma jet	Medium	Glioblastoma (U87MG)		N/A	[51]
2016	Plasma jet	Medium	Lung cancer (A549)	Smooth muscle cell (SMCs)	Yes	[34]
2016	Plasma jet	Medium	Lung cancer (A549)	Keratinocytes (HaCaT)	Yes	[34]
2016	Plasma jet	Medium	Tumor spheroid (HCT116)		N/A	[40]
2016	Others	Medium	Lung cancer (H460)	Normal lung (L132)	Yes	[19]
2016	Plasma jet	Medium	Glioblastoma (U251SP)	Breast epithelium (MCF10A)	Yes	[47]
2016	Plasma jet	PBS	Colon cancer (HCT116)	Skin fibroblasts (NHSF)	No	[14]
2016	Plasma jet	PBS	Melanoma (Lu1205)	Lung fibroblast (MRC5)	No	[14]
2016	DBD	Medium/ PBS/FBS	Cervical cancer (HeLa)	Hamster cells (CHO-K1)	Yes	[15]
2016	Plasma jet	Medium	Cervical cancer (HeLa)	Lung fibroblast (Wi38)	Yes	[35]
2016	Plasma jet	Medium	Breast cancer (MDA-MB-453)	Breast epithelium (MCF10A)	Yes	[35]
2016	Plasma jet	Medium	Colon cancer (HCT116)		N/A	[35]
2016	Plasma jet	Medium	Melanoma (A375)	Dermal fibroblasts (HDFs)	Yes	[36]
2016	Plasma jet	Medium	Lung cancer (A549)	Dermal fibroblasts (HDFs)	Yes	[36]
2016	Plasma jet	Medium	Osteosarcoma (MG63)	Melanocytes	Yes	[36]
2016	Plasma jet	Ringer's solution	Glioblastoma (U251SP)	Epithelial cell (MCF10A)	Yes	[52]
2016	Plasma jet	Ringer's solution	Glioblastoma (U251SP)	Neonatal keratinocyte	Yes	[52]
2016	Plasma jet	Medium	Lung cancer (A549)		N/A	[53]

(Continued)

Table 5.1. (*Continued*)

Years	CAP	Solutions	Cancer cell lines	Normal cell lines	Selectivity	Reference
2016	Plasma jet	Medium	Bladder cancer (SCaBER)	Epithelial cell (MDCK)	Yes	[37]
2017	Plasma jet	Medium	Hepatocellular carcinoma (HepG2)	Liver cell (L02)	Yes	[38]
2017	Plasma jet	Medium/PBS	Pancreatic cancer (PA-TU-8988T)		N/A	[16]
2017	Plasma jet	Medium/PBS	Glioblastoma (U87MG)		N/A	[16]
2017	Plasma jet	Medium/PBS	Breast cancer (MDA-MB-231)		N/A	[16]
2017	Plasma jet	Medium	Glioblastoma (U87MG)		N/A	[54]
2017	Plasma jet	Medium	Ovarian cancer (ES_2)		N/A	[55]
2017	Plasma jet	Medium	Ovarian cancer ($SKOV_3$)		N/A	[55]
2017	Plasma jet	Medium	Gastric cancer (SC-2-NU)		N/A	[56]
2017	Plasma jet	Medium	Gastric cancer (AGS)		N/A	[56]
2017	Plasma jet	Medium	Lymphoma (MWCL-1)		N/A	[57]
2017	Plasma jet	Medium	Lymphoma (RPCI-WM1)		N/A	[57]

Clearly, such selectivity is determined by the sensitivity of cells to long-lived reactive species or other formed components. Like the direct treatment, selectivity should be due to the rise of specific intracellular ROS after PAS treatment (figure 5.4) [39]. The selective anti-cancer effect of PAM has also been observed in a co-culture system with the liver cancer cell line HepG2 and normal cell line L02 [38]. In addition, the anti-cancer capacity of PAM over colon adenocarcinoma multicellular tumor spheroids (MCTS) has also been investigated *in vitro* [40]. A noticeable double-strand break (DSB) occurred at the peripheral part of MCTS [40].

5.2 Effective components in PAS

Chemical factors, particularly reactive species, contribute to the cytotoxicity of PAS. In plasma medicine, the main chemical factors to consider include the change of pH, the CAP-originated species, and the chemicals generated by the interaction between CAP-originated species and components in an aqueous solution. Buffering chemicals in PAS, such as medium, PBS, and Ringer's solution, protect PAS from non-negligible changes in pH due to CAP treatment [41, 42]. So, the CAP-triggered pH change does not contribute to the cytotoxicity of PAS. Such a pH change may be a factor for the toxicity of PAS if the buffering chemicals are removed from the solutions, or the treatment time increases in length. However, the meaning of CAP treatment and corresponding reactive species is negligible in such a situation.

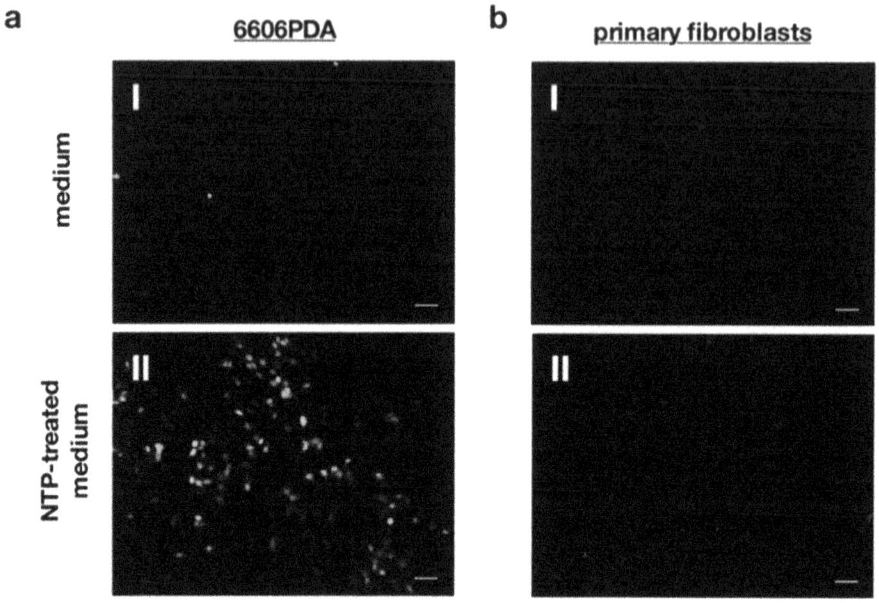

Figure 5.4. Selective rise of intracellular ROS in the PAM-treated pancreatic cancer 6606PDA cells [39]. (a) Pancreatic cell line. (b) Normal fibroblast cell line. Cells were first stained with CM-H_2DCF-DA and subsequently exposed to argon-treated medium (I) and NTP-treated (60 s) medium (II). Scale bars: 50μm. Reproduced from [39]. CC BY 4.0.

Among the diverse ROS/RNS, H_2O_2 plays an important role in the cytotoxicity of PAS [13, 28, 30, 34, 37, 40, 43–45]. NO_2^- alone will not cause observable cytotoxicity to cancer cells [16]. However, H_2O_2 alone can cause noticeable inhibition and cell death on cancer cells, though it cannot cause the same cytotoxicity as PAS [46]. Synergistically using H_2O_2/NO_2^- or $H_2O_2/NO_2^-/NO_3^-$ in medium or PBS can generate a similar selective cytotoxicity as PAS does [14, 47]. NO in PAS may also contribute to its cytotoxicity. Dissolution of NO in PAS significantly increases the cytotoxicity of PAS to HeLa cells [48]. Other species such as $ONOO^-$ may also play a key role. We screened a H_2O_2-resistant skin melanoma cell line B16F10. Surprisingly, the H_2O_2-resistant cell line could also be effectively killed by CAP-activated RPMI1640 medium. Clearly, PAS treatment cannot be regarded as a simple H_2O_2 treatment. The whole picture of the specific biological function of reactive species in PAS is far from clear.

Moreover, the reaction between CAP and components in solutions may form new toxic products or modifications on small molecules. For example, Hiromasa Tanaka, *et al* found that the lactate in Ringer's solution is necessary for the cytotoxicity of the CAP-activated Ringer's solution (PAL) on glioblastoma cells [52]. PAL cytotoxicity may be due to the CAP-triggered modification of lactate, such as the formation of acetyl and pyruvic acid-like groups [52]. Among the diverse components in PAL, 2,3-dimethyltartrate exhibited selective cytotoxicity on glioblastoma cells, glyoxylate exhibited cytotoxicity on both glioblastoma and normal

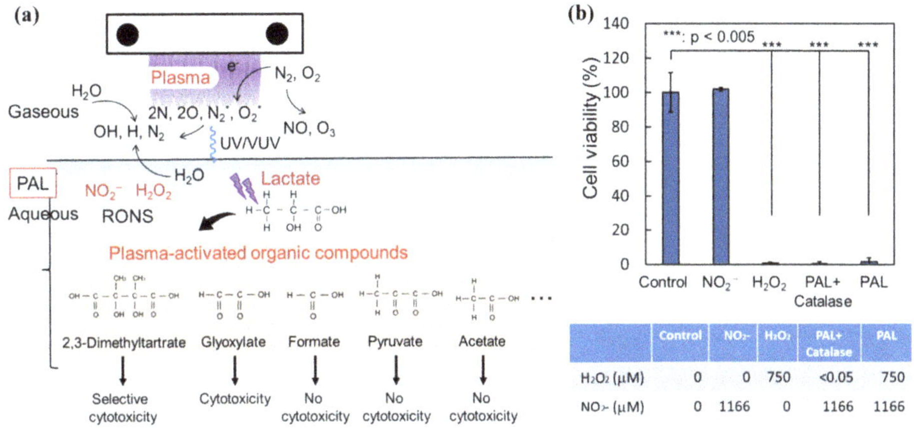

Figure 5.5. CAP-activated Ringer's lactate solution's anti-cancer effect [58]. (a) Components in PAL identified by NMR analysis and direct infusion-ESI–MS/MS analysis. (b) Cytotoxic effects of components in PAL. Reproduced ftom [58].

cells, and pyruvate, formate, and acetate did not show any cytotoxicity on glioblastoma cells (figure 5.5(a)) [58]. Even if catalase completely consumed H_2O_2 in PAL, a significant killing effect on glioblastoma U251SP cells was seen (figure 5.5(b)) [58]. Similarly, the CAP-treated fetal bovine serum (FBS) has also shown cytotoxicity over cervical cancer HeLa cells and hamster cells CHO-K1 [15]. The potential chemical modification on the contents of FBS is thought to be the main reason for such cytotoxicity [15]. However, due to the extremely long CAP treatment time length, such as 10 min in this study, the thermal deactivation of the proteins in FBS may also happen. Therefore, the underlying mechanism for the so-named cytotoxicity of the CAP-treated FBS still needs further clarification.

In addition to the reactive species mentioned above, the carrier, such as cell medium or PBS, is another key factor in determining the cytotoxicity of PAS. Despite the common understanding that the reactive species from CAP cause the death of cancer cells, it has been found that their cytotoxicity to cancer cells disappears when the PAS is made of simple buffered solutions such as PBS [33, 52]. For a long time, scholars even believed that only PAM had anti-cancer capacity. Consequently, researchers speculated that the toxic chemicals formed by the reaction between reactive species and components in the medium were the main cytotoxic actors in PAM [33]. Our recent study first demonstrated that all PAS, including CAP-activated PBS, showed noticeable cytotoxicity over cancer cells as long as the CAP treatment is adequately long [16]. Dilution of PBS in DMEM did not change H_2O_2 generation (figure 5.6(a)) but significantly weakened the cytotoxicity of PAM on breast cancer MDA-MB-231 cells and pancreatic adenocarcinoma PA-TU-8988T cells (figures 5.6(b) and (c)) [16]. However, glioblastoma U87MG cells did not follow the same trend (figure 5.6(d)) [16]. In addition, U87MG cells and PA-TU-8988T cells were specifically vulnerable to the CAP-activated PBS and PAM, respectively [16]. Specific rise of intracellular ROS in U87MG and PA-TU-8988T cells may cause different vulnerabilities of cancer cells to

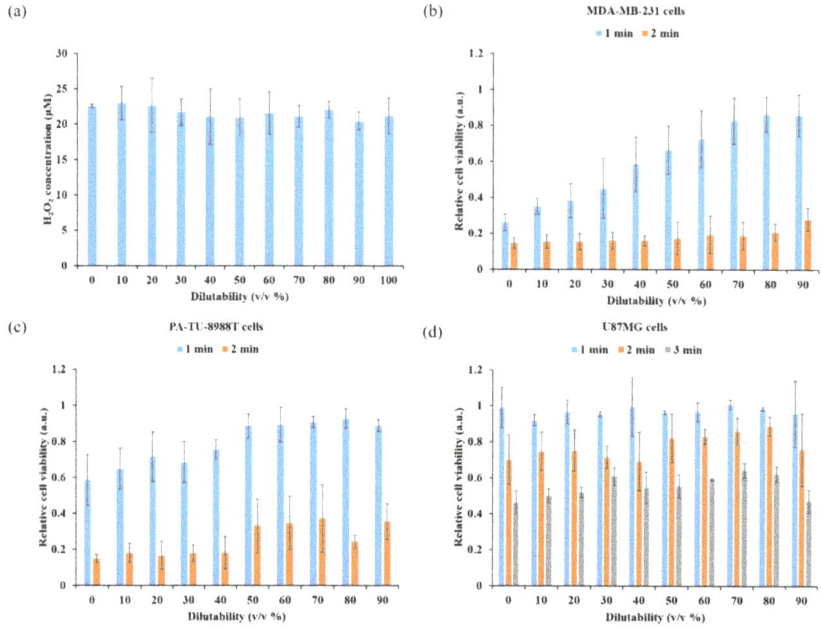

Figure 5.6. The dilution in medium by PBS significantly weakens the anti-cancer performance of PAS in some cancer cell lines [16]. (a) H_2O_2 in PAM with different dilutabilities (PBS/DMEM). (b) Breast adenocarcinoma cell line (MDA-MB-231). (c) Pancreas adenocarcinoma cell line (PA-TU-8988T). (d) Glioblastoma cell line (U87MG). 0% of dilutability and 90% of dilutability represent a mixed DMEM composed of 0% (v/v) PBS + 100% (v/v) DMEM and a mixed DMEM composed of 90% (v/v) PBS + 10% (v/v) DMEM, respectively. Reproduced from [16].

PSM and the CAP-activated PBS [16]. Similar phenomena have been observed in the PAM-treated glioblastoma U251SP cells, rather than the CAP-activated lactate (PAL) cases (figure 5.7) [52]. Specific cellular response in nutrient-starved or nutrient-rich environment may cause cancer cells to respond differently to the extracellular reactive species, such as H_2O_2 [16]. Thus, the previous observation by other scholars reflects only a superficial understanding of the specific vulnerabilities of cancer cells to PAS. For example, specific cell lines may be more vulnerable to one PAS, such as PAM, than another PAS, such as the CAP-activated PBS.

Extending the CAP treatment time is the most common strategy to increase the cytotoxicity of PAS on cancer cells [10, 14, 15, 30, 52]. We have identified three other key factors that can be regulated to enhance the cytotoxicity of PAS: the gap between the plasma source and solution, the interface area between solution and plasma, and the volume of solution [46]. Recently, these variables have been tested in other CAP-activated solutions, such as PBS by other groups [56]. Specifically, increasing the interface between the plasma and solution surface will generate a higher concentration of reactive species in PAS [30]. For example, the PAS made in a six-well plate will be significantly more toxic to cancer cells than the PAS made in a

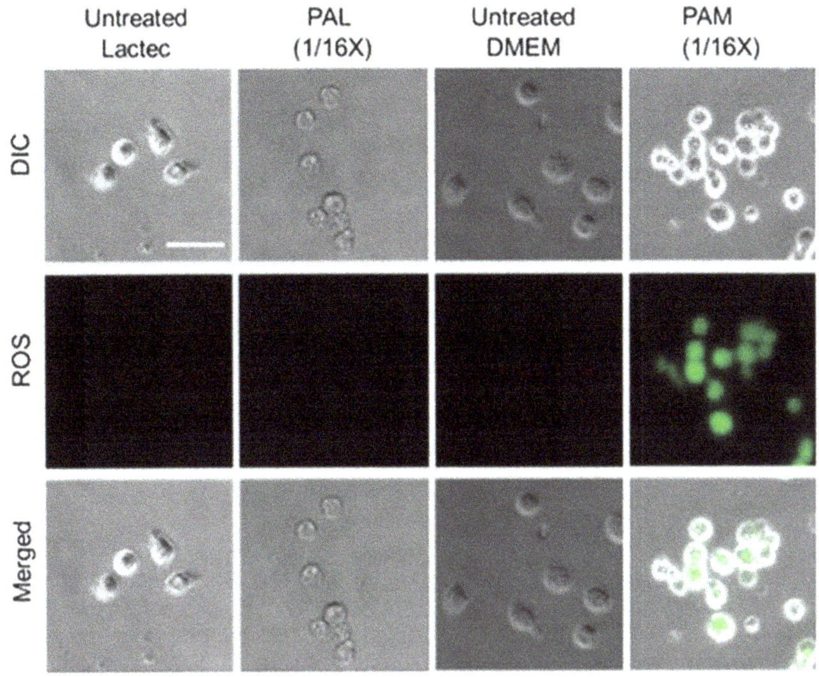

Figure 5.7. Absence of intracellular ROS rise after PAL treatment [52]. Glioblastoma U251SP cells were treated with PAL and PAM in an 8-well chamber slide. Intracellular ROS was stained by CM-H2DCFDA. Reproduced from [52]. CC BY 4.0.

96-well plate. In addition, a shorter gap between the plasma source and medium will also make a more toxic PAS with a higher reactive species concentration [30, 56]. The stronger cytotoxicity of PAS can also be achieved using a smaller volume of medium during the CAP treatment [30, 56].

Complicated reactions commonly exist at the interface between bulk plasma and aqueous solutions. These reactions may cause chemical modification of components in the CAP-treated solutions. Several chemicals are highly reactive with the species in PAS like H_2O_2. First, FBS, a necessary component of a typical complete cell culture medium, is highly reactive to many effective species, particularly ROS in PAM. FBS eliminates the cytotoxicity of PAS on cancer cells [12]. In addition, pyruvate, another component commonly found in cell culture medium, are highly reactive with the H_2O_2 in PAM [43]. Among all 20 amino acids, cysteine is most reactive with the key anti-cancer species such as H_2O_2 in PAM [12]. Based on these discoveries, an ideal solution for PAS should avoid containing FBS, pyruvate, as well as cysteine.

5.3 Instability in storage

Despite being a potential benefit in the clinical application, the stability of PAS during storage is a challenge. As a pharmacological modality, it must be stable under

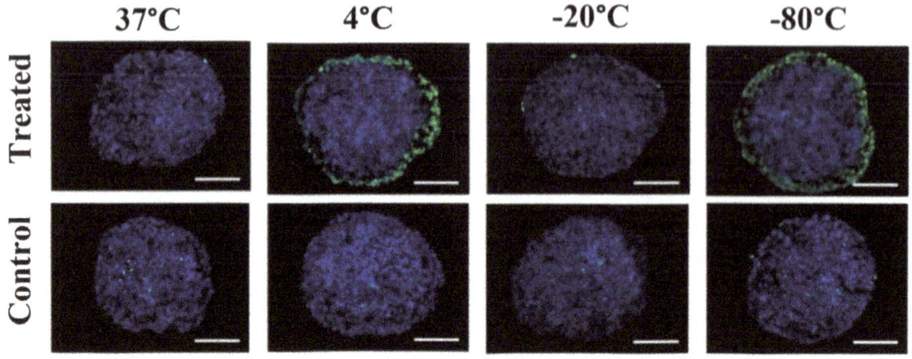

Figure 5.8. PAM storage [40]. The phosphorylation of histone variant H_2AX (γ-H_2AX, green) stained DNA damage. DAPI (blue) stained the nuclei. White scale bar: 100 μm. PAM-treated and the control 3D multicellular tumor spheroids were stained 4 h post treatment. PAM has been stored at 37 °C, 4 °C, −20 °C, and −80 °C for 7 days. Reproduced from [40]. CC BY 4.0.

relatively low-cost conditions. Liquid nitrogen provides an extremely cold temperature (−196 °C) storage condition, but comes at a high financial cost and is not practical when a large amount of PAS is needed. A more cost effective storage option would be either 2 °C–8 °C, −20 °C, or room temperature. Many studies, however, indicate that PAM must be stored at −80 °C or even −150 °C for full, long-term stability [40, 43, 52]. As early as the earliest report about the cytotoxicity of PAM on cancer cells published, the degradation of PAS during storage has been extensively observed. Both the anti-cancer species, and as a result, the anti-cancer capacity of PAM is gradually degraded during the storage over a wide temperature range [12, 30, 36, 39, 43] from room temperature and 4 °C to as low as −20 °C to −30 °C (figure 5.8) [12, 40, 43]. PAM's degradation was slightly inhibited at 2 °C–8 °C, but in general, the natural degradation of PAM/PAS at a wide temperature range may severely weaken its pharmacological value [12, 43].

The degradation mechanism of PAS provides clues to solve this dilemma. By comparing the H_2O_2 concentration in CAP-activated PBS and CAP-activated Dulbecco's Modified Eagle Medium (DMEM) after the storage at 8 °C, 22 °C, and −25 °C, we discovered that the degradation of PAM was mainly due to the reaction between the CAP-originated species and the components in medium (figure 5.9(a)) [45]. Cysteine and methionine were the main components causing the degradation of PAM [45]. CAP-activated DMEM without cysteine and methionine was much more stable than CAP-activated DMEM after storage at 8 °C, 22 °C, as well as at −25 °C (figures 5.9(b) and 5.9(c)) [45]. In short, these studies demonstrate that the degradation of PAS can be prevented in an economically feasible way if reactive components, such as cysteine and methionine, at least when such PAS is stored at around 8 °C.

Furthermore, 3-nitro-L-tyrosine (3NT) was discovered as an anti-degradation additive providing another strategy to resist the degradation of PAM at 2 °C–8 °C, that works even if cysteine and methionine are present [29]. Chemical formulas of

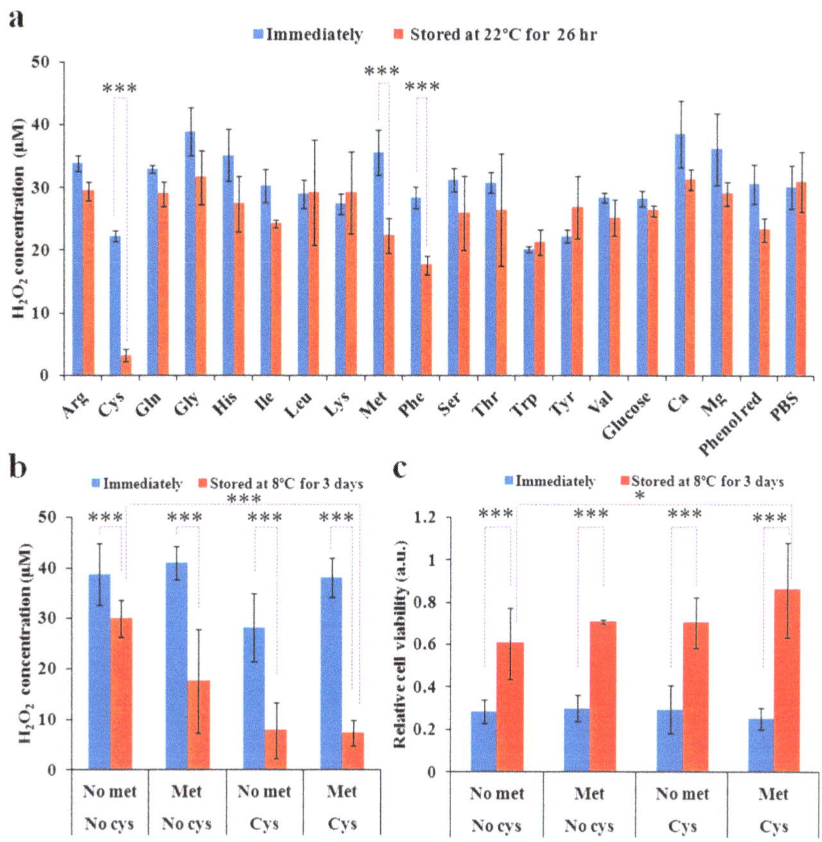

Figure 5.9. Cysteine and methionine mainly contribute to the degradation of the CAP-activated DMEM [45]. (a) H_2O_2 in the CAP-activated PBS containing specific component during the storage at 22 °C for 26 h. (b) H_2O_2 in the CAP-activated cysteine/methionine-free DMEM, cysteine-free DMEM, methionine-free DMEM, and standard DMEM during the storage at 8 °C for 3 days. (c) Anti-cancer capacity of the CAP-activated cysteine/methionine-free DMEM, cysteine-free DMEM, methionine-free DMEM, and standard DMEM during the storage at 8 °C for 3 days. Reproduced from [45]. CC BY 4.0.

phenylalanine, tyrosine, and 3NT were shown in figure 5.10(a). Among the 15 amino acids included in DMEM, only tyrosine and phenylalanine have an aromatic ring structure. The sole difference between tyrosine and phenylalanine is a hydroxyl group on the aromatic ring of tyrosine. As shown in figure 5.10(b), only 3NT-containing (2 mM) DMEM shows a strongly improved stability after storage at 8 °C for 26 h. This suggcsts that thc nitro group on thc aromatic ring of tyrosine may endow 3NT with its capacity to inhibit the degradation of H_2O_2 during the storage [45]. Furthermore, 3NT effectively inhibits the degradation of H_2O_2 in the CAP-activated cysteine-containing (0.2 mM) PBS at 22 °C for 26 h (figure 5.10(c)). Thus, PAM may be stabilized by 3NT via inhibiting the reaction between some specific components in DMEM, such as cysteine, with H_2O_2 [45].

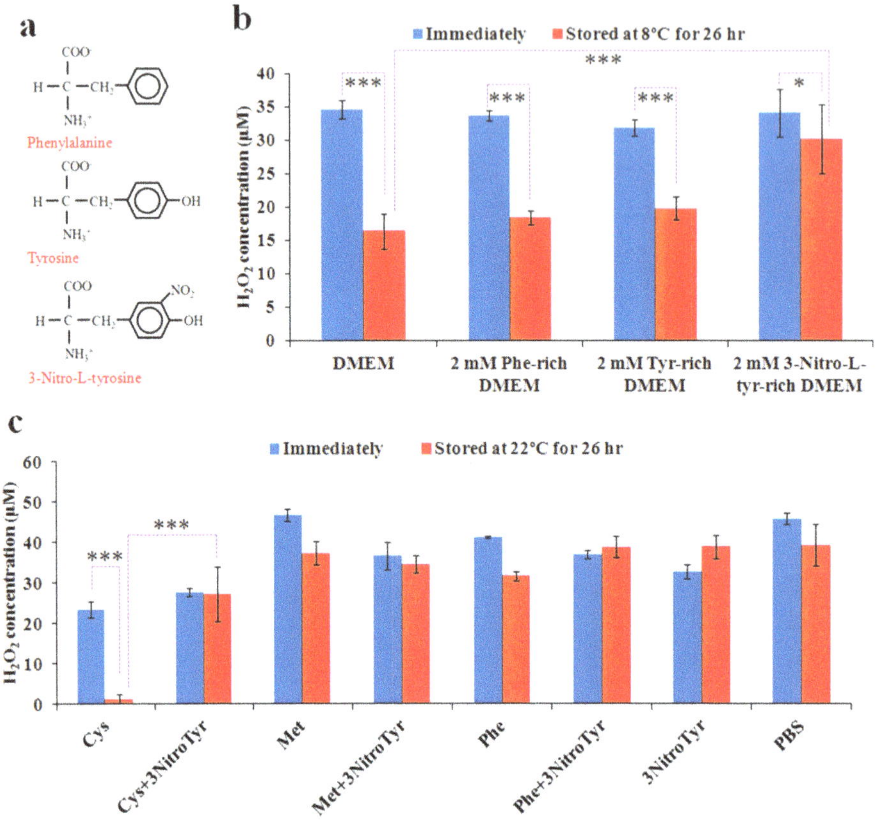

Figure 5.10. 3-Nitro-L-tyrosine (3NT) inhibits the degradation of CAP-stimulated DMEM. (a) Chemical formulas of phenylalanine, tyrosine, and 3NT [45]. (b) H_2O_2 in the CAP-activated standard DMEM, phenylalanine-containing DMEM, tyrosine-containing DMEM, and 3-Nitro-L-tyrosine-containing DMEM after the storage at 8 °C for 26 h [45]. (c) H_2O_2 in the CAP-activated single amino acids-containing PBS and double amino acids-containing PBS during the storage at 22 °C for 26 h. Concentration of cysteine, methionine, phenylalanine, and 3NT in PBS were 0.2, 0.2, 0.4 and 2mM, respectively. Reproduced from [45]. CC BY 4.0.

The concentration of 3NT in PBS is a key factor affecting its anti-degradation efficacy during storage. The anti-degradation effect of 3NT-PBS was investigated by comparing H_2O_2 in CAP-activated PBS and CAP-activated 3NT-PBS (0–9 mM) before and after storage. One can see that H_2O_2 in CAP-activated PBS decreased by about 32% after storage for 7 days at 4 °C. However, degradation trend ceased and even reversed after adding 3NT in PBS from 0.2 to 0.8 mM. When 3NT in PBS was higher than 1 mM, more H_2O_2 was generated in the CAP-activated PBS after storage for 7 days. The maximum anti-degradation efficacy can be achieved at a relatively small 3NT concentration, such as 1 mM. Besides, a high concentration of 3NT may be toxic to cells [31]. Both the anti-degradation and safety of using 3NT need to be considered in plasma medicine applications.

So far, the most attractive feature of PAS is its ability to act independent of a direct CAP source, after being made [43, 46, 59, 60]. In other words, indirect treatment can be used as a pharmacological modality. However, a clear understanding of its chemical mechanisms must be developed before doctors, regulators, and the market can accept it. To date, only long-lived reactive species such as H_2O_2, NO_2^-, NO_3^-, and $ONOO^-$ have been regarded as effective anti-cancer components in PAS [30, 61]. While these chemical components have been effective as anti-cancer agents, they cannot explain the complex and multidimensional observed cytotoxicity alone. Varying ratios of these chemical combinations such as H_2O_2/NO_2^-, or $H_2O_2/NO_2^-/NO_3^-$ can achieve similar results to those seen after an indirect CAP treatment [14, 47, 62]. In short, the uniqueness of indirect CAP treatment compared with a simple chemical component such as long-lived reactive species treatment has not been distinguished.

References

[1] Zhang X, Li M, Zhou R, Feng K and Yang S 2008 Ablation of liver cancer cells *in vitro* by a plasma needle *Appl. Phys. Lett.* **93** 2006–9

[2] Fridman G, Shereshevsky A, Jost M M, Brooks A D, Fridman A, Gutsol A, Vasilets V and Friedman G 2007 Floating electrode dielectric barrier discharge plasma in air promoting apoptotic behavior in melanoma skin cancer cell lines *Plasma Chem. Plasma Process.* **27** 163–76

[3] Chernets N, Kurpad D S, Alexeev V, Rodrigues D B and Freeman T A 2015 Reaction chemistry generated by nanosecond pulsed dielectric barrier discharge treatment is responsible for the tumor eradication in the B16 melanoma mouse model *Plasma Processes Polym.* **12** 1400–9

[4] Kang S U *et al* 2014 Nonthermal plasma induces head and neck cancer cell death: the potential involvement of mitogen-activated protein kinase-dependent mitochondrial reactive oxygen species *Cell Death Dis.* **5** e1056

[5] Kaushik N K, Kaushik N, Park D and Choi E H 2014 Altered antioxidant system stimulates dielectric barrier discharge plasma-induced cell death for solid tumor cell treatment *PLoS One* **9** e103349

[6] Gibson A R, McCarthy H O, Ali A A, O'Connell D and Graham W G 2014 Interactions of a non-thermal atmospheric pressure plasma effluent with PC-3 prostate cancer cells *Plasma Processes Polym.* **11** 1142–9

[7] Tanaka H, Mizuno M, Toyokuni S, Maruyama S, Kodera Y, Terasaki H, Adachi T, Kato M, Kikkawa F and Hori M 2015 Cancer therapy using non-thermal atmospheric pressure plasma with ultra-high electron density *Phys. Plasmas* **22** 1–6

[8] Yan D, Sherman J H and Keidar M 2017 The application of the cold atmospheric plasma-activated solutions in cancer treatment *Anticancer Agents. Med. Chem.* **18** 769–75

[9] Yan D, Malyavko A, Wang Q, Lin L and Sherman J H 2021 Cold atmospheric plasma cancer treatment: a critical review *Appl. Sci.* **11** 7757

[10] Tanaka H, Mizuno M, Ishikawa K, Nakamura K, Kajiyama H, Kano H, Kikkawa F and Hori M 2011 Plasma-activated medium selectively kills glioblastoma brain tumor cells by down-regulating a survival signaling molecule, AKT Kinase *Plasma Med.* **1** 265–77

[11] Vandamme M *et al* 2012 ROS implication in a new antitumor strategy based on non-thermal plasma *Int. J. Cancer* **130** 2185–94
[12] Yan D, Sherman J H, Cheng X, Ratovitski E, Canady J and Keidar M 2014 Controlling plasma stimulated media in cancer treatment application *Appl. Phys. Lett.* **105** 224101
[13] Boix-garriga E, Acedo P, Casadó A, Sato T, Yokoyama M and Johkura K 2001 A key inactivation factor of HeLa cell *J. Phys. D: Appl. Phys.* **44** 37
[14] Girard P-M, Arbabian A, Fleury M, Bauville G, Puech V, Dutreix M and Sousa J S 2016 Synergistic effect of H_2O_2 and NO_2 in cell death induced by cold atmospheric He plasma *Sci. Rep.* **6** 29098
[15] Boehm D, Heslin C, Cullen P J and Bourke P 2016 Cytotoxic and mutagenic potential of solutions exposed to cold atmospheric plasma *Sci. Rep.* **6** 21464
[16] Yan D, Cui H, Zhu W, Nourmohammadi N, Milberg J, Zhang L G, Sherman J H and Keidar M 2017 The specific vulnerabilities of cancer cells to the cold atmospheric plasma-stimulated solutions *Sci. Rep.* **7** 4479
[17] Miebach L, Freund E, Clemen R, Kersting S, Partecke L-I and Bekeschus S 2022 Gas plasma-oxidized sodium chloride acts via hydrogen peroxide in a model of peritoneal carcinomatosis *Proc. Natl Acad. Sci. USA* **119** e2200708119
[18] Miebach L, Freund E, Cecchini A L and Bekeschus S 2022 Conductive gas plasma treatment a tumor toxicity of Ringer's lactate solutions in a model of peritoneal carcinomatosis *Antioxidants* **11** 1439
[19] Kumar N, Park J H, Jeon S N, Park B S, Choi E H and Attri P 2016 The action of microsecond-pulsed plasma-activated media on the inactivation of human lung cancer cells *J. Phys. D: Appl. Phys.* **49** 115401
[20] Locke B R and Shih K Y 2011 Review of the methods to form hydrogen peroxide in electrical discharge plasma with liquid water *Plasma Sources Sci. Technol.* **20** 034006
[21] Graves D B 2012 The emerging role of reactive species oxygen and nitrogen species in redox biology and some implications for plasma applications to medicine and biology *J. Phys. D: Appl. Phys.* **45** 263001
[22] Bruggeman P J *et al* 2016 Plasma–liquid interactions: a review and roadmap *Plasma Sources Sci. Technol.* **25** 053002
[23] Malyavko A, Yan D, Wang Q, Klein A L, Patel K C, Sherman J H and Keidar M 2020 Cold atmospheric plasma cancer treatment, direct versus indirect approaches *Mater. Adv.* **1** 1494–505
[24] Keidar M, Yan D and Sherman J H 2019 *Cold Plasma Cancer Therapy* (Morgan & Claypool Publishers)
[25] Bauer G 2019 Cold atmospheric plasma and plasma-activated medium: antitumor cell effects with inherent synergistic potential *Plasma Med.* **9** 57–88
[26] Bauer G 2017 Signal amplification by tumor cells: clue to the understanding of the antitumor effects of cold atmospheric plasma and plasma-activated medium *IEEE Trans. Radiat. Plasma Med. Sci.* **2** 87–98
[27] Tanaka H and Ishikawa K 2017 State of the art in medical applications using non-thermal atmospheric pressure plasma *Rev. Mod. Plasma Phys.* **1** 1–89
[28] Utsumi F, Kajiyama H, Nakamura K, Tanaka H, Mizuno M, Ishikawa K, Kondo H, Kano H, Hori M and Kikkawa F 2013 Effect of indirect nonequilibrium atmospheric pressure plasma on anti-proliferative activity against chronic chemo-resistant ovarian cancer cells *in vitro* and *in vivo* *PLoS One* **8** e81576

[29] Tanaka H *et al* 2015 Plasma with high electron density and plasma-activated medium for cancer treatment *Clin. Plasma Med.* **3** 72–6

[30] Adachi T, Tanaka H, Nonomura S, Hara H, Kondo S I and Hori M 2015 Plasma-activated medium induces A549 cell injury via a spiral apoptotic cascade involving the mitochondrial-nuclear network *Free Radic. Biol. Med.* **79** 28–44

[31] Yan D, Nourmohammadi N, Milberg J and Sherman J H 2018 Guidelines for using 3-Nitro-L-Tyrosine as an antidegradation reagent of H_2O_2 in the cold atmospheric plasma-stimulated solutions *Plasma Med.* **8** 121–30

[32] Clemen R *et al* 2021 Gas plasma technology augments ovalbumin immunogenicity and OT-II T cell activation conferring tumor protection in mice *Adv. Sci.* **8** 2003395

[33] Kalghatgi S, Kelly C M, Cerchar E, Torabi B, Alekseev O, Fridman A, Friedman G and Azizkhan-Clifford J 2011 Effects of non-thermal plasma on mammalian cells *PLoS One* **6** e16270

[34] Adachi T, Nonomura S, Horiba M, Hirayama T, Kamiya T, Nagasawa H and Hara H 2016 Iron stimulates plasma-activated medium-induced A549 cell injury *Sci. Rep.* **6** 20928

[35] Nguyen N H, Park H J, Yang S S, Choi K S and Lee J-S 2016 Anti-cancer efficacy of nonthermal plasma dissolved in a liquid, liquid plasma in heterogeneous cancer cells *Sci. Rep.* **6** 29020

[36] Saito K, Asai T, Fujiwara K, Sahara J, Koguchi H, Fukuda N, Suzuki-Karasaki M, Soma M and Suzuki-Karasaki Y 2016 Tumor-selective mitochondrial network collapse induced by atmospheric gas plasma-activated medium *Oncotarget* **7** 19910

[37] Mohades S 2016 Temporal evaluation of the anti-tumor efficiency of plasma-activated media *Plasma Processes Polym.* **13** 1206–11

[38] Duan J, Lu X and He G 2017 The selective effect of plasma activated medium in an *in vitro* co-culture of liver cancer and normal cells *J. Appl. Phys.* **121** 013302

[39] Liedtke K R, Bekeschus S, Kaeding A, Hackbarth C, Kuehn J-P, Heidecke C-D, von Bernstorff W, von Woedtke T and Partecke L I 2017 Non-thermal plasma-treated solution demonstrates antitumor activity against pancreatic cancer cells *in vitro* and *in vivo Sci. Rep.* **7** 8319

[40] Judée F, Fongia C, Ducommun B, Yousfi M, Lobjois V and Merbahi N 2016 Short and long time effects of low temperature plasma activated media on 3D multicellular tumor spheroids *Sci. Rep.* **6** 21421

[41] Panngom K, Baik K Y, Nam M K, Han J H, Rhim H and Choi E H 2013 Preferential killing of human lung cancer cell lines with mitochondrial dysfunction by nonthermal dielectric barrier discharge plasma *Cell Death Dis.* **4** e642

[42] Kaushik N, Kumar N, Kim C H, Kaushik N K and Choi E H 2014 Dielectric barrier discharge plasma efficiently delivers an apoptotic response in human monocytic lymphoma *Plasma Processes Polym.* **11** 1175–87

[43] Yokoyama M, Johkura K and Sato T 2014 Gene expression responses of HeLa cells to chemical species generated by an atmospheric plasma flow *Biochem. Biophys. Res. Commun.* **450** 1266–71

[44] Yan D, Talbot A, Nourmohammadi N, Cheng X, Canady J, Sherman J and Keidar M 2015 Principles of using cold atmospheric plasma stimulated media for cancer treatment *Sci. Rep.* **5** 18339

[45] Yan D, Nourmohammadi N, Bian K, Murad F, Sherman J H and Keidar M 2016 Stabilizing the cold plasma-stimulated medium by regulating medium's composition *Sci. Rep.* **6** 26016

[46] Yang H, Lu R, Xian Y, Gan L, Lu X and Yang X 2015 Effects of atmospheric pressure cold plasma on human hepatocarcinoma cell and its 5-fluorouracil resistant cell line *Phys. Plasmas* **22** 122006

[47] Mohades S, Laroussi M, Sears J, Barekzi N and Razavi H 2015 Evaluation of the effects of a plasma activated medium on cancer cells *Phys. Plasmas* **22** 122006

[48] Yan D, Nourmohammadi N, Talbot A, Sherman J H and Keidar M 2016 The strong anti-glioblastoma capacity of the plasma-stimulated lysine-rich medium *J. Phys. D: Appl. Phys.* **49** 274001

[49] Kurake N, Tanaka H, Ishikawa K, Kondo T, Sekine M, Nakamura K, Kajiyama H, Kikkawa F, Mizuno M and Hori M 2016 Cell survival of glioblastoma grown in medium containing hydrogen peroxide and/or nitrite, or in plasma-activated medium *Arch. Biochem. Biophys.* **605** 102–8

[50] Yan D, Xiao H, Zhu W, Nourmohammadi N, Zhang L G, Bian K and Keidar M 2017 The role of aquaporins in the anti-glioblastoma capacity of the cold plasma-stimulated medium *J. Phys. D: Appl. Phys.* **50** 055401

[51] Adachi T, Kano A, Nonomura S, Kamiya T and Hara H 2016 Histone deacetylase inhibitors stimulate the susceptibility of A549 cells to a plasma-activated medium treatment *Arch. Biochem. Biophys.* **606** 120–7

[52] Tanaka H, Nakamura K, Mizuno M, Ishikawa K and Takeda K 2016 Non-thermal atmospheric pressure plasma activates lactate in Ringer' s solution for anti-tumor effects *Sci. Rep.* **6** 36282

[53] Nakamura K, Peng Y, Utsumi F, Tanaka H and Mizuno M 2017 Novel Intraperitoneal treatment with non-thermal plasma- activated medium inhibits metastatic potential of ovarian cancer cells *Sci. Rep.* **7** 6085

[54] Takeda S *et al* 2017 Intraperitoneal administration of plasma-activated medium: proposal of a novel treatment option for peritoneal metastasis from gastric cancer *Ann. Surg. Oncol.* **24** 1188–94

[55] Wada N, Ikeda J, Tanaka H, Sakakita H, Hori M, Ikehara Y and Morii E 2017 Effect of plasma-activated medium on the decrease of tumorigenic population in lymphoma *Pathol. Res. Pract.* **213** 773–7

[56] López-Lázaro M 2006 Dual role of hydrogen in cancer: possible relevance to cancer chemoprevention and therapy *Cancer. Lett.* **252** 1–8

[57] Li Y, Ho Kang M, Sup Uhm H, Joon Lee G, Ha Choi E and Han I 2017 Effects of atmospheric-pressure non-thermal bio-compatible plasma and plasma activated nitric oxide water on cervical cancer cells *Sci. Rep.* **7** 45781

[58] Tanaka H *et al* 2021 Low temperature plasma irradiation products of sodium lactate solution that induce cell death on U251SP glioblastoma cells were identified *Sci. Rep.* **11** 18488

[59] Freund E, Liedtke K, Miebach L, Wende K, Heidecke A, Kaushik N K, Choi E H, Partecke L and Bekeschus S 2020 Plasma-activated ringer's lactate solution displays a selective cytotoxic effect on ovarian cancer cells *Cancers* **12** 476

[60] Bekeschus S, Kolata J, Winterbourn C, Kramer A, Turner R, Weltmann K D, Br B and Masur K 2014 Hydrogen peroxide: a central player in physical plasma-induced oxidative stress in human blood cells *Free Radic. Res.* **48** 542–9

[61] Sato T, Yokoyama M and Johkura K 2011 A key inactivation factor of HeLa cell viability by a plasma flow *J. Phys. D: Appl. Phys.* **44** 372001

[62] Xu D, Cui Q, Xu Y, Liu Z, Chen Z, Xia W and Kong M G 2018 NO_2^- and NO_3^- enhance cold atmospheric plasma induced cancer cell death by generation of $ONOO^-$ *AIP Adv.* **8** 105219

Chapter 6

Physics based cold atmospheric plasma therapy

As an ionized gas from a non-equilibrium discharge process, cold atmospheric plasma (CAP) acts through chemical and physical factors. Previously, only reactive species and other chemical factors were regarded as the main players in plasma medicine. Due to lacking direct experimental evidence, physical effect has been ignored for decades. Until very recently, via novel experimental designs, the strong physically-based anti-cancer effect has been revealed in several cell lines. In chapter 6, we give a comprehensive introduction to the research status and some preliminary discussion on mechanisms and significance of physically-based CAP treatment as a non-invasive approach, which may have a deep impact on future plasma medicine research.

6.1 Limitation of chemically-based treatment

The conventional approach and founding vision of CAP as a source of reactive species limits its application range. The efficacy of CAP treatment has widely been regarded to come directly from its chemical factors, particularly reactive species. Reactive species impacts a variety of cells, including bacterial, plant, and mammalian cells, particularly cancer cells, in a multitude of ways [1, 2]. However, under the chemical paradigm, if a cell line is resistant to damage by reactive species, it will also resist CAP treatment. For example, we found that B16F10 melanoma cells were very resistant to reactive species *in vitro*, particularly ROS such as H_2O_2 [3]. This reactive species-resistance may arise from the combination of many complex mechanisms, at least including strong intracellular antioxidant system, robust reactive species transmembrane diffusion, and high-fidelity genome repair processes [4]. Obviously, without novel approaches to explore the potential of CAP sources, biological resistance to chemical species will remain a natural barrier for plasma medicine.

doi:10.1088/978-0-7503-5537-7ch6

Conventional chemically-based treatment generates reactive species that chiefly leads to apoptosis of mammalian cells. CAP-triggered apoptosis follows well-understood apoptotic pathways with characteristic cellular changes, such as cell shrinkage, DNA fragmentation, membrane budding, and formation of final apoptotic bodies with membranous vesicles [5, 6]. Flow cytometric data reveals the molecular markers to describe and quantify the apoptotic rate after treatment. Additionally, necrosis has been observed in flow cytometry data, but its causal mechanism is unknown [7–10]. Apoptosis does not trigger an immune or inflammatory response [11–14]. Instead, necrosis which involves the bulk leakage of solutions containing cellular components, will trigger an immune response *in vivo*.

6.2 Revealing physical effect

Recently, CAP treatment based on physical factors has been reported to conquer the sensitivity limitation of cancer cells to reactive species. This approach differs from previous studies because it is able to cause significant anti-cancer effects in reactive species-resistant cell lines via a brand-new underlying mechanism. Currently, this concept and approach has begun to attract more and more attention in academia. Here, we present a preliminary introduction on the research status and basic discoveries from the latest publications.

To demonstrate the physically-based biological effect, we need to harness novel strategies. Because chemical and physical factors co-exist in CAP, the complete blockage of chemical factors is the prerequisite. The results of these studies challenged the previous notion that direct contact with reactive species was indispensable to cause the biological effects associated with CAP treatment. In our earliest trials, we treated the bottom of an inverted (upside-down) multi-well plate or cell culture dish, which allowed us to easily observe a strong physically-based effect. Overnight supernatant was removed first before treatment. Due to the water surface tension, there was still a thin medium layer to cover cells, particularly at the junction between wall and bottom. As a control, we tested and found that cells left in this inverted setting for 20 min did not impact cell viability [15–17]. After treatment, cells were immediately returned to standard culture conditions with new medium (figures 6.1(a) and (b)). For the widely used polystyrene or polyene multi-well plates or dishes, the bottom thickness or lid thickness was ~1 mm. Interestingly, the physical factors in CAP affected an area beyond the size of the plasma plume. This prompted us to present the cell viability data of the multi-well plate in a new way: 3D cell viability maps (figure 6.1(c)). Cell viability was measured using the Thiazolyl Blue Tetrazolium Bromide (MTT) assay. Values were normalized by the dividing experimental group by the control group [16]. For the 96-well plate, 3D cell viability maps visualized the normalized viability of the middle 10 × 6 wells, which reflected the physical effect's spatial distribution on cells.

To observe cellular responses using optical microscopes, some general guidelines can be given here. As shown in figure 6.2(a), a treatment was performed on the bottom of an inverted (upside-down) dish (100 mm). APPJ's plasma plume's tip had

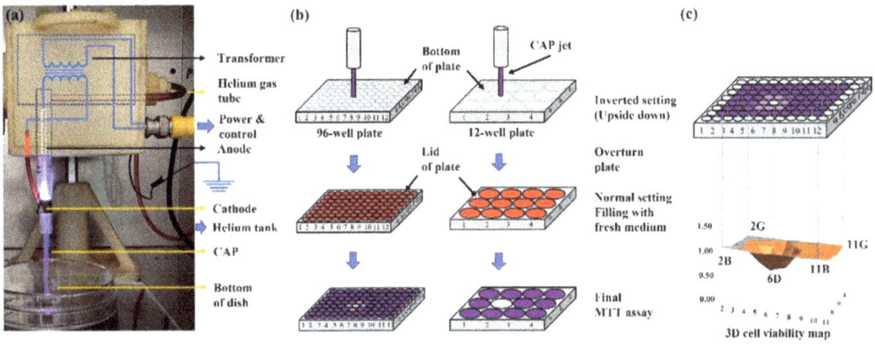

Figure 6.1. APPJ source and physically-based treatment. (a) APPJ source. (b) Basic strategies to perform chemically and physically-based treatment on a 96-well or 12-well plate. (c) 3D cell viability map of a 96-well plate, presenting all middle 10 ×6 wells' cell viability simultaneously. Five marked well numbers represented the boundary wells and the treated '6D' well. Reproduced from [18]. CC BY 4.0.

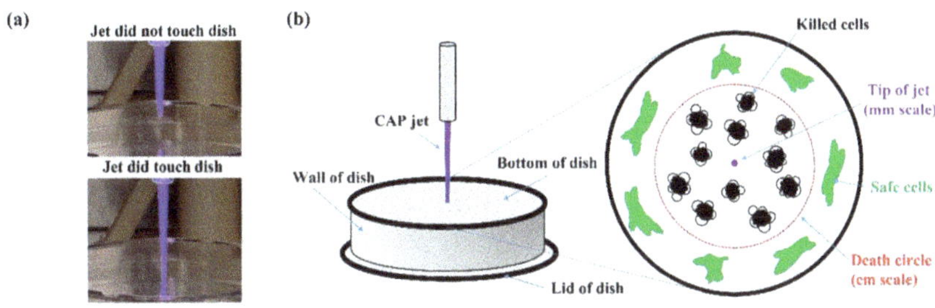

Figure 6.2. Cell death circle in a physically-based treatment. (a) APPJ in operation. (b) Schematic illustration of microscopic observation on cell death. 'Death circle' is much larger than the plasma plume's tip. Reproduced from [18]. CC BY 4.0.

a diameter smaller than 1 mm when it did not contact the dish bottom. Even plasma plume's tip touched the dish bottom, the contacting area diameter stayed on the scale of millimeters range. Notably, despite having a small area in direct contact with the dish, the plasma plume affected a much bigger area on the scale of centimeters (figure 6.2(b)). Most physically damaged or killed cancer cells were in a circular area named the 'death circle'. Cells located outside the 'death circle' remained safe. In chapter 6, many conclusions are based on investigations in such a 'death circle'.

6.3 Necrotic cell death

Skin cancer is one of the most common global malignancies. Most skin cancers are classified as either melanoma or non-melanoma. Although non-melanoma skin cancers such as basal and squamous cell carcinomas are more common in the population, melanomas are the deadliest [19–21]. Currently, despite the many treatment options for melanoma, novel non-invasive therapeutic methods are still in need [22–24]. Our previous studies found that B16F10 melanoma cells are very resistant to ROS, such as H_2O_2 *in vitro* [3]. In other words, chemically-based treatment did not show promising results on B16F10 cell lines. Finding an alternative way to use CAP in melanoma therapy is a critical step to demonstrate the feasibility of CAP as a cancer treatment.

We first demonstrated the strong anti-cancer effect on B16F10 melanoma cells caused by physical factors in CAP. We compared the growth inhibition of chemically-based and physically-based treatment on B16F10 cells. Due to the strong resistance to ROS, chemically-based treatment did not cause noticeable inhibition on growth B16F10 cells. However, physically-based treatment caused a much stronger growth inhibition. Three minutes of chemically-based treatment caused very minimal growth inhibition on B16F10 cells. In contrast, just 1.5 min of physically-based CAP treatment caused ~60% growth inhibition (figures 6.3(a)–(c)). This is the critical advantage of physically-based treatment: strong killing effect on reactive species-resistant cells.

Different anti-cancer performance is due to the different cell death mechanisms. Microscopic bright field imaging was performed on cells 1- and 2-days post treatment and we observed changes not reported in previous studies. These cellular changes were characterized by the cytoplasmic aggregation toward the nucleus and

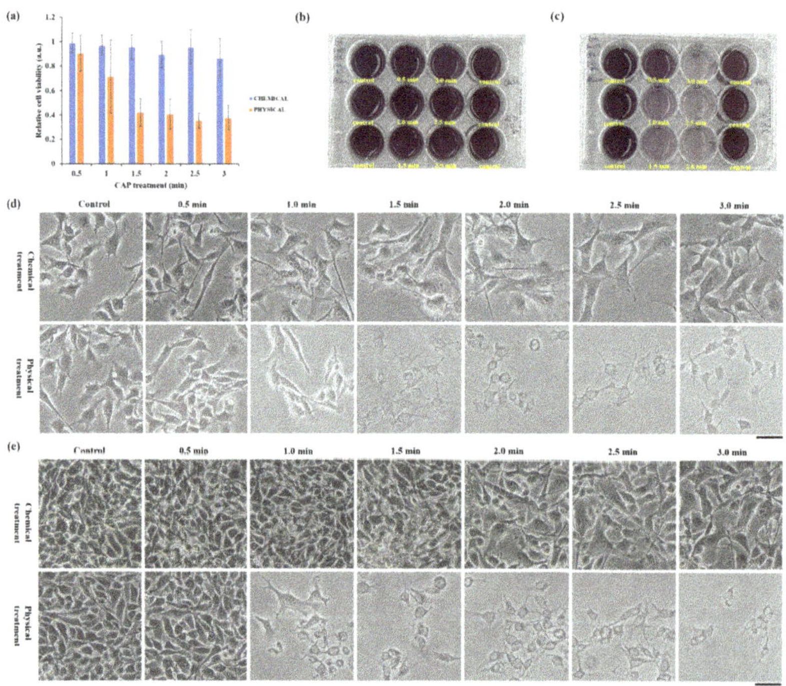

Figure 6.3. Different cellular responses to chemically and physically-based treatment [16]. (a) MTT cell viability assay. (b and c) MTT assay after chemically and physically-based treatment. (d and e) Bright field imaging 1 and 2 days after treatment (0.5–3 min). Black scale bar: 50 μm. Reprinted with permission from [16], copyright (2020) American Chemical Society.

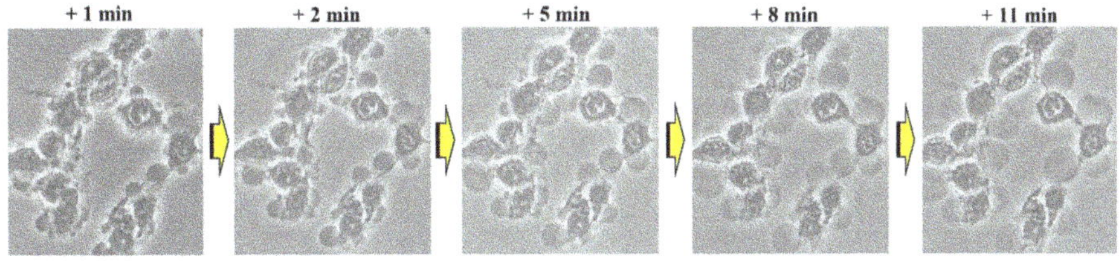

Figure 6.4. Bright field imaging of bubbling growth on the treated B16F10 cells [18]. Yellow arrows marked the bubbles on cellular membranes. '+*x* min': photos were taken at x min after treatment. Black scale bar: 50 μm. Reproduced from [18]. CC BY 4.0.

the rapid leakage of bulk water or cytosol solutions from the cytoplasmic membrane, forming a visible bubbling on cellular surface (figures 6.3(d) and (e)). These features remained for several days following the treatment, without any cell duplication, movement, or interaction. Cells seem like an 'empty shell' in such a 'fixed' state. All these features are clearly necrotic cell death features.

To further investigate the evolution of necrotic processes, we observed the change immediately after a physically-based treatment of 2 min (figure 6.4). At the very beginning, a significant shrinkage of the cytoplasm could be observed. Bubbling occurred immediately after the cytoplasmic aggregation. In the nucleus, black dots, maybe nucleoli, became more apparent, though the nucleus size did not show observable change. Bubbling growth was the main feature starting just one minute after treatment. Typically, the complete growth of a bubble took 8–11 min. As they grew bigger, the bubbles became lighter and more transparent, suggesting a slow dilution of the internal components [16]. Bubbles disappeared from the cellular membrane a few hours after treatment, probably after detaching. Many small, detached bubbles were found in the extracellular space after treatment [16].

While bright field phase-contrast imaging provides morphological change of treated cells, live-cell fluorescent imaging can provide more information at the cellular chemistry level. DNA and microtubules were stained with Hoechst 33342 and Green Microtubule Cytoskeleton Dye, respectively. Bubbling was clearly observed in the live fluorescent imaging using a confocal microscope (figure 6.5). The bubbles' light green surface was due to the reflection of fluorescence from microtubules. There was no fluorescence within these bubbles. We proposed that these bubbles were due to the leakage of water or cytosol components. The formation of new bubbles from a single site on the cellular membrane may suggest that they originate from membrane holes [16]. Filamentous microtubules features were lost in the treated cells. Instead, the cytoskeleton shrank and aggregated around the nucleus. Staining of DNA in the nucleus suggests that damage may happen immediately after treatment. Recently, similar features were also observed in the CAP-treated glioblastoma U87MG cells. In this study, triple staining of cell membrane (red), microtubules (green), and DNA (blue) were performed. All three fluorescence sources would project on bubbles. However, only red fluorescence appeared on the bubbles, which confirmed that each bubble had a cell membrane (figure 6.6).

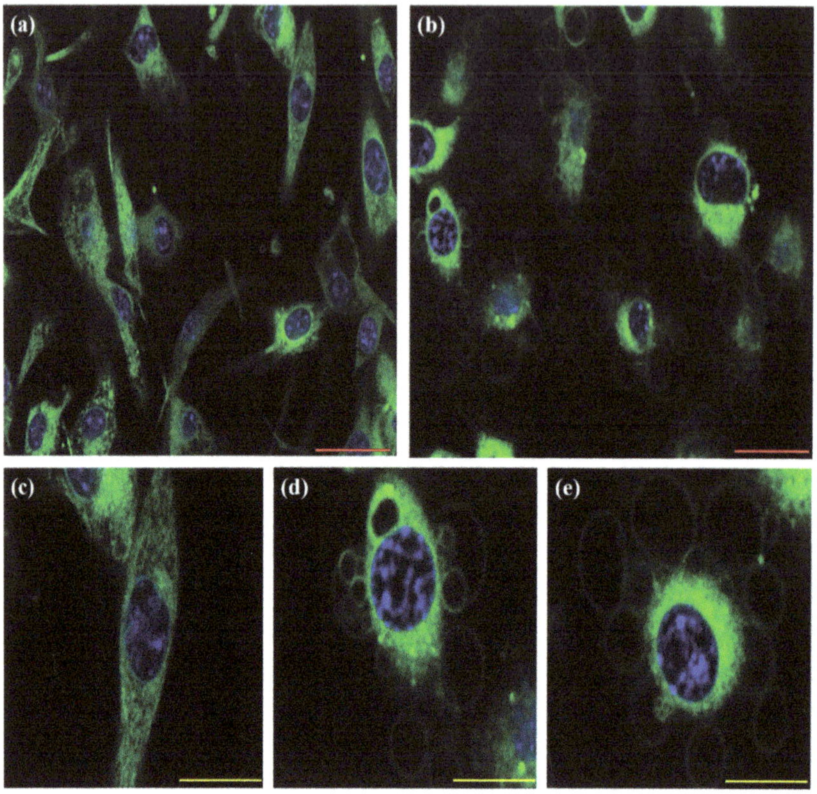

Figure 6.5. The live fluorescent imaging at 20 min after treatment [16]. (a) Control group. (b) 20 min after treatment. (c) A zoomed-in image in control group. (d) Aggregated bubbling on a single cell. (e) A single cell is surrounded by cell-size bubbles. Microtubules and DNA were stained in green and blue, respectively. Scale bars: 40 μm (red), 20 μm (yellow). Reprinted with permission from [16], copyright (2020) American Chemical Society.

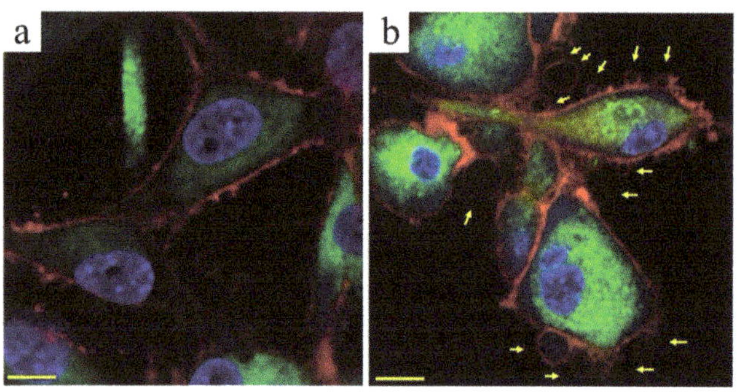

Figure 6.6. Triple staining of the treated glioblastoma U87MG cells [25]. (a) Control group. (b) 10 min post 60 s of treatment. Yellow arrows marked bubbles on cellular membranes. Cell membrane, microtubules, and DNA were stained in red, green, and blue, respectively. Yellow scale bar: 10 μm. Reprinted with permission from [25], copyright (2023) American Chemical Society.

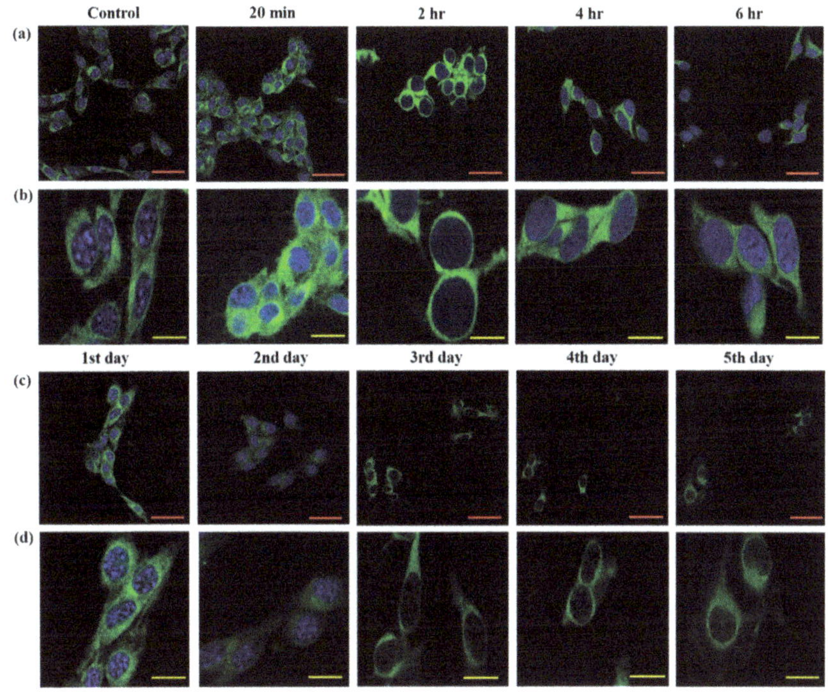

Figure 6.7. Fluorescent imaging of the treated B16F10 cells over a long timescale [16]. (a) Cells were observed 20 min, 2, 4, and 6 h after the treatment. (b) The zoomed-in images. (c) Cells were observed 1, 2, 3, 4, and 5 days after the treatment. (d) The zoomed-in images. Microtubules and DNA (nucleus) were stained in green and blue, respectively. Red/yellow scale bars: 40/20 μm. Reprinted with permission from [16], copyright (2020) American Chemical Society.

Confocal imaging was further used to study the evolution of treated cells over a long-time range. Several trends were found, consistent with the observations from bright field imaging. A summary is given here based on figure 6.7. First, bubbles could not be seen on cells at two hours post treatment. The nucleus expanded in the initial 2 h and seemed to reach a maximum size. After that, the nucleus experienced a slight decrease in size. Second, over the 5-day observation, a gradual loss of DNA in the nucleus was seen. Blue DNA staining was obvious 1 day post treatment, however, the signal was weak at day 2 and went completely dark 3 days post treatment, suggesting a complete DNA loss at that time point [16]. Third, treated cells appeared 'fixed' over the whole process. Green microtubule staining also became weak starting on the second day after treatment. At 5 days after treatment, the appearance of B16F10 cells was an empty, circular shell composed of microtubules, other cytoskeletal components, and other cytosolic components not stained here [16]. These features explain the 'fixed' shape of treated cells observed in bright field imaging. By the sixth day post treatment, no cellular remnants were found in imaging.

To date, studies reporting necrosis from CAP treatment based their conclusions mainly on flow cytometry. This study is the first demonstration of necrotic features

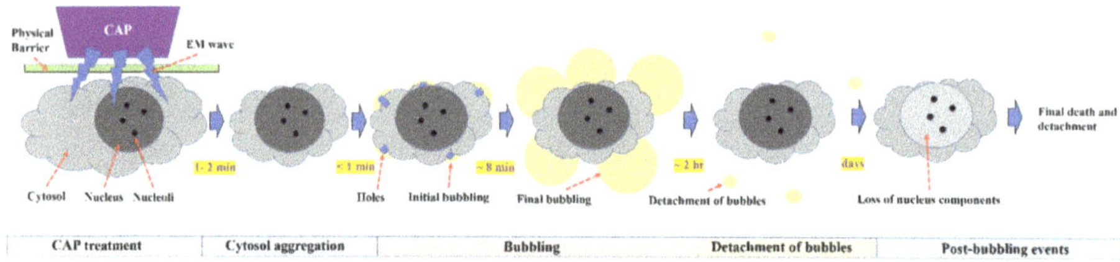

Figure 6.8. Schematic illustration of physically-caused necrotic process. Necrosis has main four stages: cytosol aggregation, bubbling, detachment of bubbles, as well as post-bubbling events. Reprinted with permission from [16], copyright (2020) American Chemical Society.

characterized by cytosol aggregation and subsequent bulk leakage of water out of cells (figure 6.8). Completely different from apoptotic processes, these necrotic processes are seen just a couple of minutes post treatment. Based on microscopic imaging, these bubbling phenomena are due to hole formation on the cytoplasmic membrane. In the long-term, cells gradually lose DNA from the nucleus in the days post treatment. The aggregated cytoskeleton stays intact for 3–5 days and finally forms an empty shell, with a 'fixed shape' that is maintained for days until the final detachment or absence from the substrate [16]. However, we need to point out that bubbling phenomena can also be observed in other cell death process such as pyroptosis [26].

6.4 Bubbling and osmotic pressure

Intracellular osmotic pressure across the cell membrane plays an important role in regulating cellular shape. As we introduced above, strong cytoplasm aggregation may create intracellular pressure and even pores on the cell membrane, which may be the mechanism responsible for the bubbling phenomena during physically-based treatment. Deionized (DI) water is a hypotonic solution that causes the water influx into cells and can potentially counteract the intracellular pressure and bubbling on the cell membrane. Given this rationale, we studied the cellular response in a series of media (RPMI)/DI (Milli-Q) diluted with water after physically-based treatment. After 3 min of treatment, B16F10 cells were immediately replenished with different RPMI/Milli-Q mixed solutions. Bubble density on cells decreased as the volume ratio of DI water increased in the mixed solutions (figure 6.9). When the CAP-treated cells were immersed in 10 mL of DI water, the bubbling phenomena was completely counteracted. Due to the water influx, even the CAP-treated cells also expanded in DI water compared to the cells in RPMI.

Normally, DI water exerts a large osmotic pressure on mammalian cells, which may cause cytolysis after cells' swelling. We found different cellular responses to the osmotic pressure with and without physically-based treatment (figure 6.10). The untreated B16F10 cells experienced a typical mammalian cellular response in a hypotonic environment. They gradually swelled, formed spheres, detached from the substrate, and finally burst. Debris was left on the substrate which, looked like the nucleus remained after the cytolysis, suggesting that the nucleus has much stronger

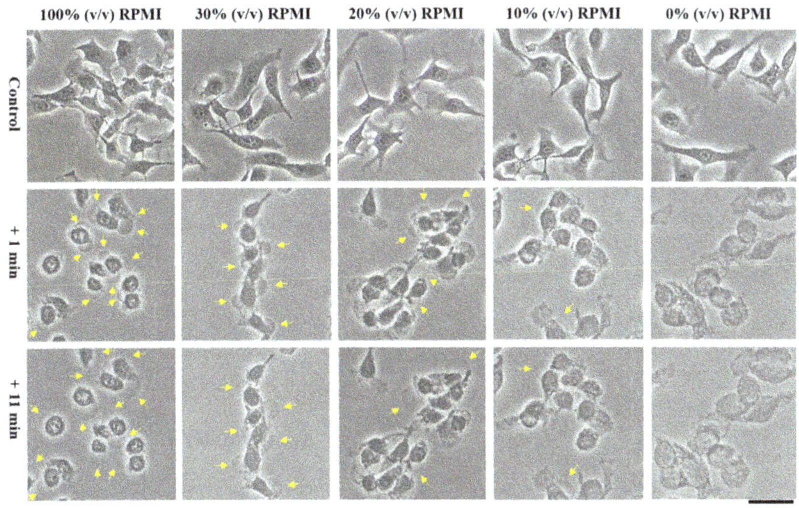

Figure 6.9. Hypotonic solutions inhibit the bubbling phenomena on the CAP-treated melanoma cells [16]. The volume ratio % (v/v) of RPMI in the mixed solution (RPMI/DI water) was 100%, 30%, 20%, 10%, and 0%, respectively. '+*x* min': images were taken at *x* min after treatment. Bubbles were marked by yellow arrows. Black scale bar: 50 μm. Reprinted with permission from [16], copyright (2020) American Chemical Society.

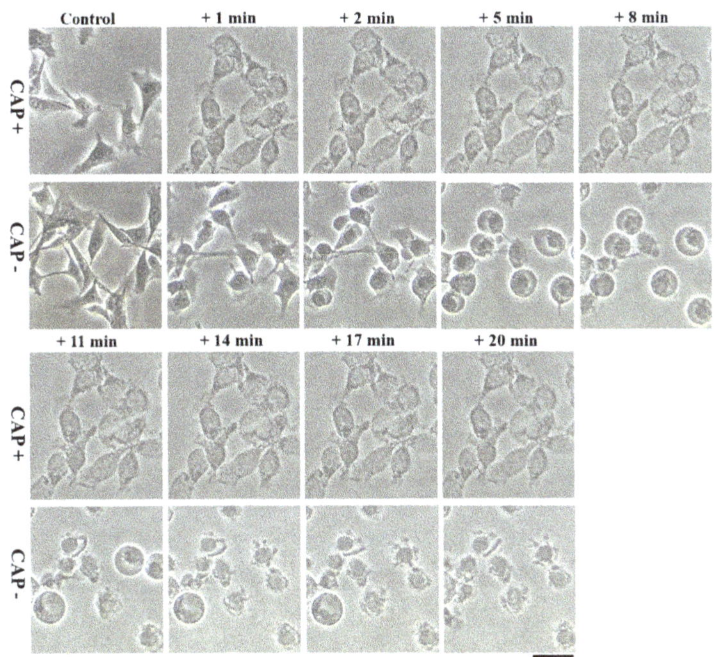

Figure 6.10. Physically-based treatment changed cellular osmotic responses in DI water [16]. '+*x* min': images were taken at x min after cells were immersed in DI water. 'CAP+'/'CAP−': with and without CAP treatment, respectively. Black scale bar: 50 μm. Reprinted with permission from [16], copyright (2020) American Chemical Society.

mechanical strength than the cell membrane and cytoskeleton networks. In contrast, the CAP-treated B16F10 cells did not show similar changes during the 20 min imaging process. Our fluorescent imaging revealed that the nucleus did not change significantly after physically-based treatment; the nucleus seemed like a 'fixed cellular shape' or a 'tough shell'. The main effect from CAP was cytosol aggregation. Therefore, it is possible that the treated cells used their nucleus to resist osmotic pressure: cells already lose most cytosol and the integrity of cytoskeleton system. Obviously, the nucleus shows a much stronger mechanical strength than the cell membrane. Therefore, the CAP-treated cells did not show typical cellular response to a large osmotic pressure.

6.5 Anti-cancer physical factors in CAP

Physically-based treatment is different from previous studies not only because of the stronger anti-cancer efficacy in many cell lines, but also because of the novel underlying mechanism. What causes these physically-based biological effects? Typically, CAP will generate at least three physical effects: thermal effect, ultra-violet (UV) effect, as well as electromagnetic (EM) effect. To understand the role of individual physical factors, we first analyzed the thermal effect of CAP. The heating effect on the inverted 'upside-down' 35 mm glass-bottom dish was measured after a 2 min CAP treatment. The maximum temperature was 38.5 °C ± 0.4 °C [16]. To investigate the cellular response in a similar heating environment, B16F10 cells in a 35 mm glass-bottom dish were partially immersed in a water bath for 2 min (figure 6.11), which had a temperature of 41 °C–43 °C [16]. To simulate physically-based CAP treatment, bulk medium was removed before the heating treatment. The new medium was quickly renewed after treatment. Heating tests did not cause any change in cellular morphology [16]. This was the earliest demonstration that the heating effect alone was not likely to cause the CAP's physical effect on cancer cells.

UV and EM effects were further investigated. A UV reflection film was used to block UV radiation from plasma, which showed an excellent attenuation over 191–400 nm [16]. As shown in figure 6.12, the UV reflection film was pasted on the 'upside-down' 96-well plate during CAP treatment. This film was removed after treatment. The two sizes of UV reflection film covered 5 × 5 and 10 × 6 wells on a 96-well plate. To investigate EM effect, we used a copper sheet as an EM effect blocker also with two sizes (5 × 5 wells, 10 × 6 wells). The copper sheet was set between the APPJ source and the top of an inverted 96-well plate during treatment. We found that an 8 min physically-based treatment caused strong growth inhibition on B16F10 cells in the presence of a UV reflection film, regardless of sizes [16]. In contrast, the 5 × 5 well-size copper sheet already strongly counteracted the killing effect. The 10 × 6 well copper sheet completely protected B16F10 cells from the physically-based killing effect. Clearly, EM rather than UV effect determined the anti-melanoma performance.

Similar studies were carried out on glioblastoma U87MG cells and similar trends were observed. Copper sheets with different sizes were put above a heating reflection

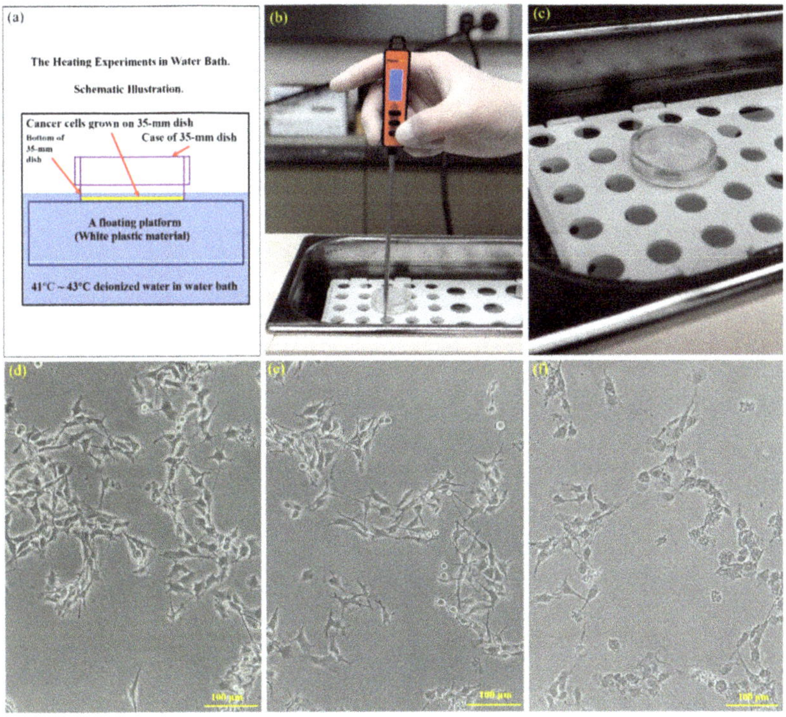

Figure 6.11. Heating experiment in the water bath [16]. (a) Schematic illustration. The dish bottom was immersed in the water with a temperature 41 °C–43 °C. The medium was removed before the experiment. (b) Heating (41.0 °C). experiment in performance. (c) 35 mm dish on a white floating flatform. (d–f) Phase-contrast microscopic imaging of the control group, the cells after a 2 min of heating test, and the cells after 2 min of physically-based treatment. Images were taken 11 min post treatment. Reprinted with permission from [16], copyright (2020) American Chemical Society.

film during each treatment [15]. A 1 × 1 well- size copper sheet did not weaken the physically-based killing effect. A 2 × 2 well-size copper sheet showed a weak inhibition on the killing effect. When a copper sheet covered more wells, a much stronger inhibition of the physical effect was observed. The physically-based killing effect of 8 min CAP treatment could be completely inhibited when a copper sheet covered 5 × 5 wells [15]. Given these observations, a reasonable design of a metal pad may guide the physical effect only towards tumorous sites while protecting normal tissues by covering them. Together, these studies indicate that the physically-based killing effect is likely due to the EM effect rather than the UV or thermal effect from CAP.

6.6 Modulating physical effects

In the conventional view of plasma medicine, the efficacy of chemically-based treatment is solely determined by the chemical factors, particularly the reactive species concentration. For physically-based treatment, its efficacy will only be

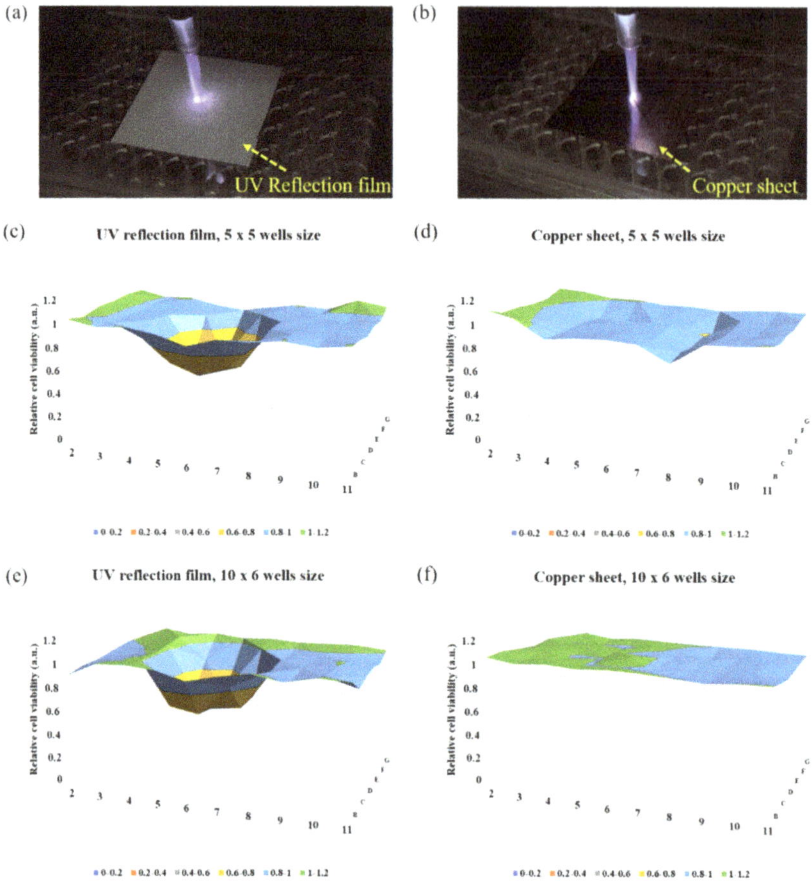

Figure 6.12. The anti-melanoma effects can be blocked by a copper sheet, but not a UV reflection film [16]. (a and b) The treatments on the bottom of inverted 96-well plates covered by UV reflection and copper sheet. Here, UV reflection film and the copper sheet had a 5 × 5 wells size. (c–f) 3D cell viability. Reprinted with permission from [16], copyright (2020) American Chemical Society.

determined by physical factors. Bubbling is the most representative necrotic feature after physically-based treatment. Different from conventional chemically-based treatment, the physically-based treatment does not require direct contact of bulk plasma with the target or the diffusion of reactive species to the target. However, a suitable range of operational parameters is also necessary to observe physically-based anti-cancer effect, just like what we have learned in chemically-based treatment [27–29].

The gap between the plasma plume and the target is a critical parameter determining the efficacy of physically-based treatment. Here, we demonstrated that the plasma plume caused strong bubbling phenomena on melanoma B16F10 cells even though there was an air gap between bulk plasma's tip and target. We investigated the bubbling phenomena at different gap lengths between the APPJ's nozzle and the dish's back surface (figure 6.13). Neither too small nor too long a gap

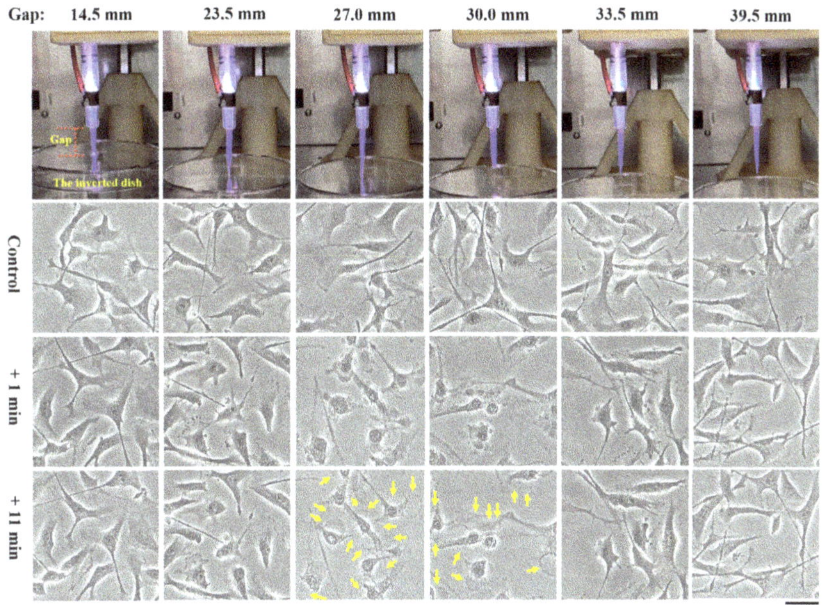

Figure 6.13. Bubbling phenomena occurred on B16F10 cells when APPJ's nozzle separated a moderate gap from the surface of an inverted dish [16]. The experimental group images were taken *in situ*. '+*x* min': images were taken at *x* min after treatment. Some bubbles were marked by yellow arrows. Black scale bar: 50 μm. Reprinted with permission from [16], copyright (2020) American Chemical Society.

would trigger bubbling and necrosis [16]. Even when the plasma plume's tip noticeably contacted the upside-down dish's surface (gap = 14.5 mm), the physical effect did not cause observable cellular change. A large gap (33.5–39.5 mm) between the plasma plume's tip and the back surface of the dish did not cause physical impact either. Bubbling phenomena only occurred when there was a moderate gap. Here, when the plasma plume tip just touched the surface (gap = 27.0 mm), a strong bubbling phenomenon was observed. Similarly, bubbling phenomenon could be observed when there was an air gap between the plasma plume' tip and surface (gap = 30.0 mm).

The flow rate of helium gas is another critical operational parameter to modulate the physical effect on cancer cells. It is well known that a moderate flow rate tends to generate maximum reactive species, such as H_2O_2 [30–32]. Interestingly, the impact of helium flow rate on the physical effect shared a similar trend. When helium flow rate increased from 0.28 to 2.34 lpm, cells did not show a proportional response. Noticeable necrotic features occurred only when the flow rate was at middle range, from 0.67 to 1.25 lpm (figure 6.14(a)). In contrast, neither too small nor too large a flow rate would trigger these necrotic features. Cell viability assays supported a similar trend quantitatively. A strong growth inhibition effect was observed only at intermediate helium flow rate range, from 0.98 to 3.33 lpm (figure 6.14b). It seems that specific experimental conditions such as the dish size, might affect the physical effect without changing the general trend. For example, the maximum growth

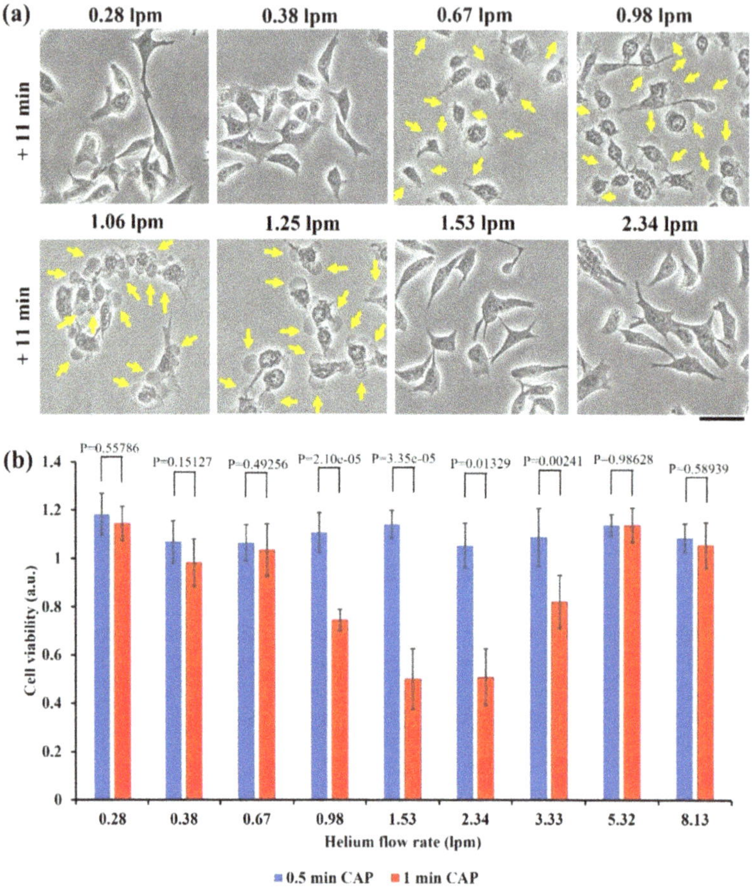

Figure 6.14. Helium flow rate affects physically triggered necrosis in B16F10 cells [18]. (a) Morphological changes in 100 mm dish after the treatment (3 min) with different flow rates. Yellow arrows marked bubbles on cellular membranes. '+11 min': phase-contrast imaging was taken at 11 min post treatment. Black scale bar: 50 μm. (b) Cell viability assay in 12-well plates. Reproduced from [18]. CC BY 4.0.

inhibition occurred when flow rate was 1.53 or 2.34 lpm, which was different from the treatment performed on a 100 mm dish.

6.7 Trans barrier capability

The advantages of physically-based treatment do not lie solely in the stronger anti-cancer performance compared with chemically-based treatment. An important benefit that chemically-based treatment does not have is the trans barrier capability, which may realize the non-invasive application of plasma medicine. Cancer is a large threat to human health [33]. The exploration for topical, local, and non-invasive therapeutic approaches in the management of cancer is a mainstay in medicine, particularly for superficial tumors such as melanoma and squamous cell carcinoma [33]. Regarding melanoma, due to its cutaneous or mucosal nature, it naturally becomes the most promising candidate for the CAP-based cancer therapy [33]. Nonetheless, the tumorous tissue in melanoma is not directly exposed to plasma without surgery. There is no need to mention other deeper tumor cases. It is urgent to explore the non-invasive potential of CAP, particularly in a contactless treatment route.

For *in vitro* studies, cell culture dishes and multi-well plates are made of plastic materials such as polystyrene. These materials are not penetrable to all chemical factors in CAP. Above studies already demonstrated that the EM effect could penetrate the bottom of multi-well plates or dishes with a thickness around 1 mm to kill cancer cells [15–17]. A question then arises: what is the penetration depth of the physical effect in CAP when it faces polystyrene materials? Recently, we found a 40% growth inhibition on B16F10 cells after 2 min of physically-based treatment with a 6 mm thick glass barrier and 1 mm thick polystyrene material between plasma plume and cells (figure 6.15). Typical necrotic features could be observed in bright field imaging, and they were consistent with the trends revealed by the cell viability MTT assay [18]. Necrotic and cell death features were counteracted when seven glass slides were used as the barrier.

Physically triggered cell death is not only observed when the treatment is performed in an inverted 'upside-down' configuration [18]. In a new demonstration,

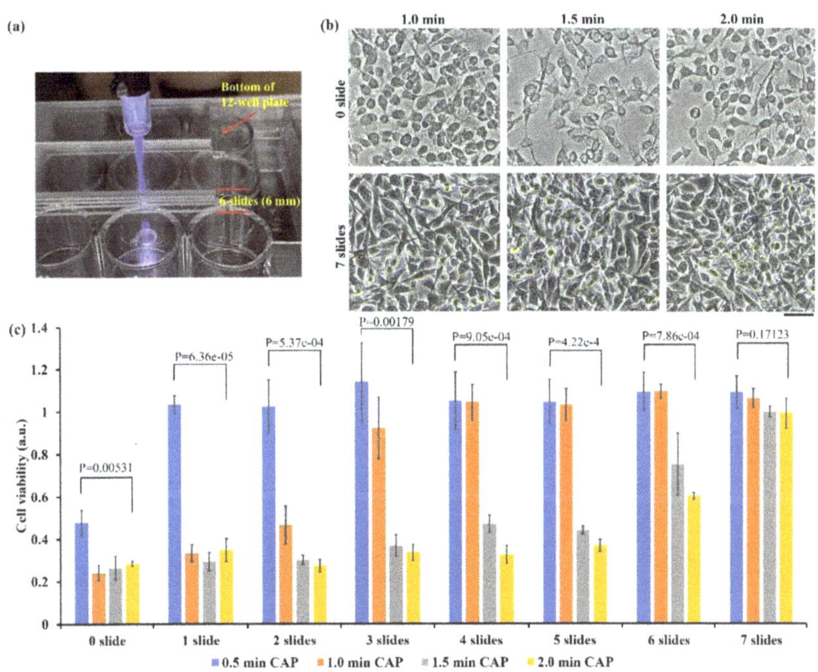

Figure 6.15. Penetration of EM effect through several glass layers to affect B16F10 cells [18]. (a) A representative treatment on 6 glass slides (6 mm) above the inverted 12-well plate. (b) Representative bright field images 1 day after treatment. '0 slide': treatment was performed without glass slides. Black scale bar: 50 μm. (c) MTT cell viability assays. Reproduced from [18]. CC BY 4.0.

CAP treatment was performed in a normal 'right side-up' configuration. The natural sensitivity of EM effect to the distance may determine the effective working range of the physical effect in CAP. Strong physically triggered necrosis was observed when treatment was performed on the bottom of an inverted 'upside-down' slide (figure 6.16(a)). When treatment was performed on the lid of a 'right side-up' configuration, the necrosis did not occur as well, because there was a large (13 mm) air gap between plasma and the target (figure 6.16(b)). However, as the air gap between the lid and the cells continued to decrease to 5 or 2 mm, a sign of strong necrosis finally appeared (figures 6.16(c) and (d)). Therefore, the distance between the plasma source and its target rather than the specific test configuration determines the physically-based biological effect.

Water may be a crucial factor affecting the biological impact of CAP treatment, particularly when physical factors need to play the dominant role. Water may block the EM effect, particularly at mm-wavelength range. Here, we demonstrated that a water layer completely blocked the physically-based killing effect in an

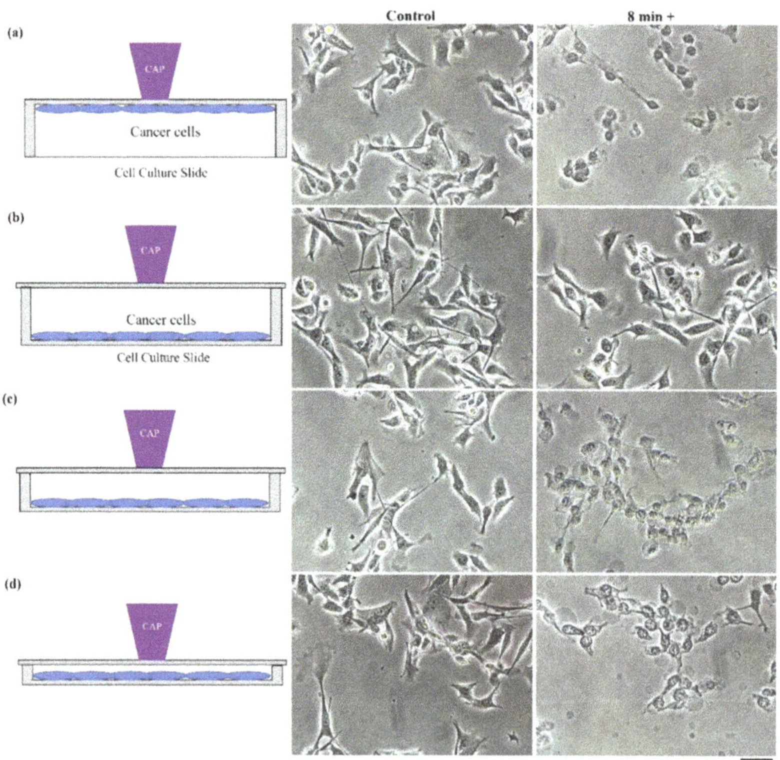

Figure 6.16. The gap between plasma plume' tip and cells is crucial for anti-melanoma effect [18]. The bulk medium was removed during treatment and quickly renewed after treatment. '8 min+': images were taken 8 min after a 2 min of treatment. (a) Treatment on an 'upside-down' cell culture chamber slide (gap = 1 mm). (b–d) Treatment on the lid of a 'right side up' cell culture chamber slides with a gap of 13, 5, and 2 mm. Black scale bar: 50 μm. Reproduced from [18]. CC BY 4.0.

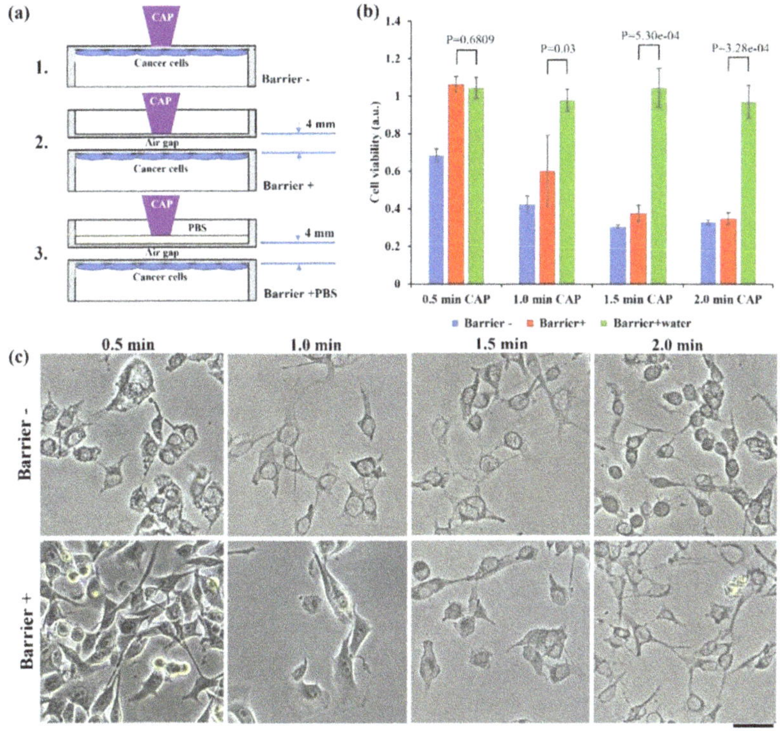

Figure 6.17. Water layer blocked physically-based anti-melanoma effect [18]. (a) Experimental design. (b) Water layer completely inhibited anti-melanoma effect. (c) Bright field imaging 1 day after treatment. Black scale bar: 50 μm. Reproduced from [18]. CC BY 4.0.

inverted 'upside-down' setting. As shown in figure 6.17, both polystyrene material and an air gap (2 mm) did not block the killing effect. However, the killing effect was weakened when these physical barriers became thicker. However, a phosphate buffered saline (PBS) layer completely inhibited this physically-based killing effect. Considering the relatively dry nature of skin versus the medium environment during *in vitro* studies, physical factors in CAP may play important roles in animal studies, particularly in subcutaneous tumor cases, which have a relatively close distance to bulk plasma.

Physical factors in CAP not only cause a direct but contactless killing effect on cancer cells, they also provide novel ways to extend the application of plasma medicine. In addition to these APPJ source-based physical treatments, a newly developed CAP source, a helium gas discharge tube, provides a novel strategy to realize physically-based treatment. A helium gas discharge tube is a tunable EM emission source, allowing only plasma's physical factors to affect cells because all chemical factors are blocked in the tube [34]. Glioblastoma is one of the most aggressive and fatal brain cancers. Currently, chemotherapy is the standard modality for glioblastoma treatment and management. We demonstrated that using a helium gas discharge tube as a tunable EM emission source could sensitize

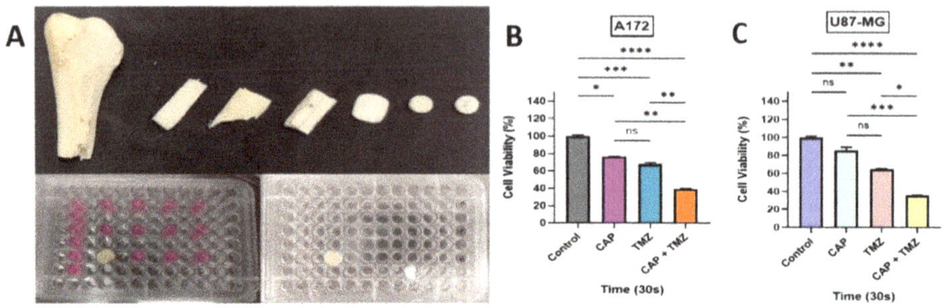

Figure 6.18. CAP can sensitize glioblastoma cells to TMZ by passing through bone barrier [36]. (a) Human fibula pieces and their positions in experiments. (b and c) Growth inhibition effects of 30 s treatment on glioblastoma cell lines: A172, and U87MG. Like previous activation studies, all cell lines were treated through blockage of bone section without the coverage of bulk media. Reproduced from [36]. CC BY 4.0.

glioblastoma cells to the cytotoxicity of a widely used drug temozolomide (TMZ) [34]. After a single 7 min treatment, the killing effect of TMZ was enhanced in U87MG and A172 cells without affecting normal human astrocytes hTERT/E6/E7. The helium discharge tube is a non-invasive device, providing a safe, stable, controllable, and low-cost routine to selectively improve current chemotherapeutic efficacy, which is particularly suitable for in-depth tumors like glioblastoma [35].

Physical factors in CAP not only penetrate glass and polystyrene but also bones. One of our recent studies found that even if pieces of human fibula were used as the barrier between the plasma plume of APPJ and the targets, two glioblastoma cell lines were still sensitized to the cytotoxicity of TMZ (figure 6.18). Like in previous studies, the fibular piece was placed on top of specific wells on a multi-well plate as a barrier. Bulk medium was removed but quickly renewed post treatment [36]. For A172 cells and U87MG cells, 30 s CAP treatment only caused ~20% growth inhibition. However, the treated cells did have the enhanced sensitivity to TMZ, showing an increase in growth inhibition from ~40% to 70%. In short, studies demonstrate the potential of using CAP source to enhance the therapeutic efficacy of existing chemotherapy in a non-invasive approach.

6.8 Physical versus chemical treatment

Some general trends of the CAP-based killing effect can be summarized here, which are not only observed in cancer treatment, but also in other CAP-treated biological samples, including mammalian cells, bacteria and even viruses. First, a direct treatment tends to cause stronger killing effect on the target than an indirect treatment with the same reactive species generation. Here, indirect treatment refers to CAP-activated solutions, such as PAM and PAS. Second, direct treatment exerts complex biological impact, which is due to long/short-lived reactive species and physical factors. For cancer cells, long-lived reactive species may be the key cytotoxic factors *in vitro* [37, 38]. However, long-lived reactive species alone cannot completely explain the observed anti-cancer efficacy *in vitro*, much less *in vivo*.

Third, short-lived reactive species and physical factors may contribute to some unique cellular responses of cancer cells [39–41], such as micromolar level cell-based H_2O_2 generation [42, 43] and activation phenomenon during direct treatment [44]. Together, these unique features cause the stronger cytotoxicity of direct over indirect treatment.

Furthermore, physical factors may play a critical role in necrosis or structural damage to cells and even viruses. In previous studies, the physical factor-based biological effects were not validated experimentally on mammalian cells. Recently, physically triggered necrosis on melanoma cells and glioblastoma cells was first demonstrated, which may be due to the EM effect from CAP. Given these observations, we further speculate that clear structural damage to bacteria and virus should be at least partially due to the similar physical effects from CAP (figure 6.19). For cancer cells, the presence or absence of an aqueous layer surrounding cells plays an important role in deciding whether chemical or physical factors have a dominating role. Physical factors may play the dominant role when the aqueous layer is adequately thin. Once a sufficient aqueous layer is introduced, physical factors are blocked, and reactive species will dominate CAP's biological effect. Here, we need to emphasize that the culture method for mammalian cells and bacteria are fundamentally different. When using agar solid medium, a widely used substrate in bacterial culture, bacteria grow above the solid medium to get nutrients from it. There is no aqueous layer to cover these bacteria when a CAP treatment is performed. Thus, it is reasonable to speculate that physical factors may be the dominant factor in these cases. The traditional understanding in plasma medicine limits potential hygiene application of CAP, which should be able to exert a strong biological effect even on a surface or interface lacking an aqueous layer.

Among these physical factors, the near-room-temperature nature of CAP and its weak UV radiation are not likely to cause observable physical destruction in both mammalian cells and bacteria. For example, the heating treatment with the same temperature of plasma plume' tip did not trigger noticeable cellular changes [16].

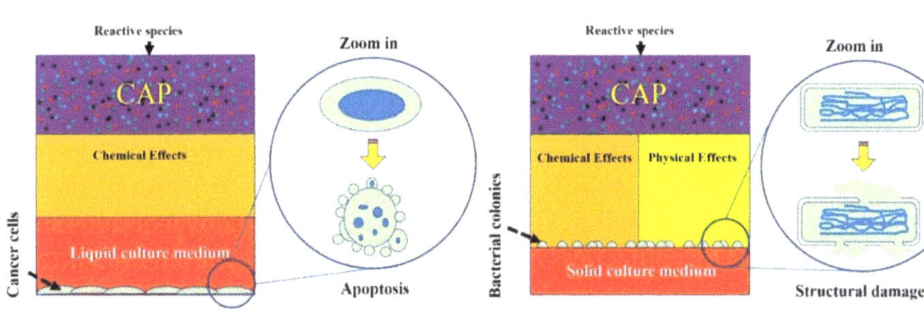

Figure 6.19. A schematic illustration of typical CAP treatment on cancer cells and bacteria *in vitro*. In most experimental settings, cancer or mammalian cells were immersed in a layer of solutions or medium during direct treatment. Such a setting naturally filters the physical effectors in CAP. In contrast, bacterial cells were directly exposed to bulk plasma because solid culture medium was widely used in bacterial culture. Reproduced from [45]. CC BY 4.0.

Higher temperature (60 °C–70 °C) on the bottom of the dish will cause strong detachment of cells. Besides, physical blockage of UV irradiation did not change the necrosis in melanoma cells. In previous studies, UV in CAP also did not kill bacterial cells grown on a solid medium [46]. Given these experimental observations, the EM effect from CAP may mainly be attributed to the necrosis in cancer cells observed so far. Furthermore, for reactive species-triggered cell death, intracellular ROS rise is the main death mechanism. Pre-treatment of cells with intracellular ROS scavengers, such as N-Acetyl-Cysteine (NAC), can effectively counteract reactive species-caused cellular responses, including intracellular ROS increase, DNA damage, mitochondrial damage, and the cytotoxicity of CAP treatment [8, 10, 47–51]. However, NAC pre-treatment did not effectively inhibit the anti-melanoma capability of physically-based CAP treatment [18].

Compared with chemically-based treatment, physically-based treatment has unique features and advantages. Physically-based treatment uses different pathways compared with reactive species-based treatment. Traditional chemically-based treatment's efficacy is based on the reactive species' concentration and cellular resistance to reactive species. For reactive species-resistant cells, the chemically-based approach does not show improved therapeutic outcomes, as in the case of melanoma cell B16F10 [3]. Promisingly, nearly all studied cell lines show a stronger sensitivity to physically-based treatment, except bladder cancer cell line MT49. These cell lines include melanoma cell line B16F10, glioblastoma cell line U87MG, lung cancer cell line A549, breast cancer cell line MDA-MB-231, and bladder cancer cell line MBT2.

Furthermore, physically-based treatment has trans barrier killing capacity. This feature provides a new perspective to understand the widely observed non-invasive anti-cancer capacity, particularly in subcutaneous tumors' response to a direct treatment [52, 53]. This capacity also sheds light on a critical challenge in plasma medicine: the long-range delivery of anti-cancer factors such as reactive species across biological barriers [54]. Currently, it is still an open question whether reactive species can cross the skin or other epithelial tissue or even bone such as skull. EM effect from CAP may penetrate some dielectric materials over a relatively long range, which may also occur *in vivo*. In our recent studies, we have demonstrated that the EM effect from CAP could kill B16F10 cells even after it passed a 6 mm thick glass and 1 mm thick polystyrene material [18]. Even though there was a 2 mm air gap and 1 mm thick of polystyrene material as the barrier, the physical effect of CAP was still observed [18]. These preliminary studies also provide critical clues to design next generation CAP sources, because glass and other dielectric materials have been widely used in current CAP sources, particularly in novel designs such as discharge tubes [34]. The contactless anti-cancer effect should get renewed attention from the community in plasma medicine because this may be the first step to use CAP in deep tumorous tissue treatment.

The efficacy of physically-based treatment may be influenced by the power and space distribution of EM emission and the relative position of the CAP source and the target. First, physical factors in CAP can affect a much larger area than the contacting area between the plasma plume tip and the target [45]. For example, the size of APPJ's plasma plume tip was roughly close to the diameter of a single well on a 96-well plate. We found that at least ten surrounding wells of the treated well experienced noticeable growth inhibition, despite all wells being physically separated. The same chemically-based treatment only caused growth inhibition in just a single well. Second, the physically-based anti-cancer performance can only be observed when plasma plumes are at a suitable relative position to the target. In some cases, the direct contact of plasma plume with the surface of the barrier (bottom of plate or dish) did not trigger cell death and observable growth inhibition. More attractively, a strong growth inhibition still occurred when the plasma plume detached from the target with a macroscale such as 8 mm. The EM emission from the plasma plume may cause these features.

More importantly, physically-based treatment builds a reasonable connection between *in vitro* and *in vivo* studies. In animal studies, a direct treatment involves the synergistic impact of physical and chemical factors on tissues, particularly when the treatment is performed on relatively dry skin. Previous studies only found apoptosis as the main cell death form in chemically-based treatment. Apoptosis will not trigger an immune or inflammation response *in vivo*. However, a series of animal studies did observe noticeable immune response [11–14], which is systematically discussed in chapter 8. Clearly, the conventional reactive species-based perspective cannot explain these dilemmas. Physically-based treatment tends to cause necrosis, involving strong leakage of cellular components into extracellular environment, an ideal condition to cause immune or inflammation response *in vivo*.

References

[1] Graves D B 2014 The emerging role of reactive oxygen and nitrogen species in redox biology and some implications for plasma applications to medicine and biology *J. Phys. D: Appl. Phys.* **45** 263001

[2] Keidar M 2015 Plasma for cancer treatment *Plasma Sources Sci. Technol.* **24** 033001

[3] Yan D, Lin L, Sherman J H, Canady J, Trink B and Keidar M 2018 The correlation between the cytotoxicity of cold atmospheric plasma and the extracellular H_2O_2-scavenging rate *IEEE Trans. Radiat. Plasma Med. Sci.* **2** 618–23

[4] Yan D, Horkowitz A, Wang Q and Keidar M 2021 On the selective killing of cold atmospheric plasma cancer treatment: status and beyond *Plasma Processes Polym.* **18** 2100020

[5] Keidar M, Shashurin A, Volotskova O, Ann Stepp M, Srinivasan P, Sandler A and Trink B 2013 Cold atmospheric plasma in cancer therapy *Phys. Plasmas* **20** 057101

[6] Yan D, Sherman J H and Keidar M 2016 Cold atmospheric plasma, a novel promising anti-cancer treatment modality *Oncotarget* **8** 15977

[7] Kim S J, Chung T H, Bae S H and Leem S H 2010 Induction of apoptosis in human breast cancer cells by a pulsed atmospheric pressure plasma jet *Appl. Phys. Lett.* **97** 023702

[8] Ahn H J, Kim K I, Kim G, Moon E, Yang S S and Lee J S 2011 Atmospheric-pressure plasma jet induces apoptosis involving mitochondria via generation of free radicals *PLoS One* **6** e28154

[9] Panngom K, Baik K Y, Nam M K, Han J H, Rhim H and Choi E H 2013 Preferential killing of human lung cancer cell lines with mitochondrial dysfunction by nonthermal dielectric barrier discharge plasma *Cell Death Dis.* **4** e642

[10] Ahn H J, Kim K I, Hoan N N, Kim C H, Moon E, Choi K S, Yang S S and Lee J S 2014 Targeting cancer cells with reactive oxygen and nitrogen species generated by atmospheric-pressure air plasma *PLoS One* **9** e86173

[11] Mizuno K, Yonetamari K, Shirakawa Y, Akiyama T and Ono R 2017 Anti-tumor immune response induced by nanosecond pulsed streamer discharge in mice *J. Phys. D: Appl. Phys.* **50** 12LT01

[12] Lin A, Truong B, Patel S, Kaushik N, Choi E H, Fridman G, Fridman A and Miller V 2017 Nanosecond-pulsed dbd plasma-generated reactive oxygen species trigger immunogenic cell death in A549 lung carcinoma cells through intracellular oxidative stress *Int. J. Mol. Sci.* **18** 966

[13] Lin A, Gorbanev Y, Cos P, Smits E and Bogaerts A 2018 Plasma elicits immunogenic death in melanoma cells regulation of antigen-presenting machinery in melanoma after plasma treatment *Clin. Plasma Med.* **9** 9

[14] Lin A, Gorbanev Y, De Backer J, Van Loenhout J, Van Boxem W, Lemière F, Cos P, Dewilde S, Smits E and Bogaerts A 2019 Non-thermal plasma as a unique delivery system of short-lived reactive oxygen and nitrogen species for immunogenic cell death in melanoma cells *Adv. Sci.* **6** 1802062

[15] Yan D, Wang Q, Malyavko A, Zolotukhin D B, Adhikari M, Sherman J H and Keidar M 2020 The anti-glioblastoma effect of cold atmospheric plasma treatment: physical pathway v.s. chemical pathway *Sci. Rep.* **10** 11788

[16] Yan D, Wang Q, Adhikari M, Malyavko A, Lin L, Zolotukhin D B, Yao X, Kirschner M, Sherman J H and Keidar M 2020 A physically triggered cell death via transbarrier cold atmospheric plasma cancer treatment *ACS Appl. Mater. Interfaces* **12** 34548–63

[17] Wang Q, Malyavko A, Yan D, Lamanna O K, Hsieh M H, Sherman J and Keidar M 2020 A comparative study of cold atmospheric plasma treatment, chemical versus physical strategy *J. Phys. D: Appl. Phys.* **54** 095207

[18] Yan D, Wang Q, Yao X and Malyavko A 2021 Anti-melanoma capability of contactless cold atmospheric plasma treatment *Int. J. Mol. Sci.* **22** 11728

[19] Miller A J and Mihm M C 2006 Melanoma *N. Engl. J. Med.* **355** 51–65

[20] Thompson J F, Scolyer R A and Kefford R F 2005 Cutaneous melanoma *Lancet* **365** 687–701

[21] Linares M A, Zakaria A and Nizran P 2015 Skin cancer *Primary Care: Clin. Off. Pract.* **42** 645–59

[22] Soengas M S and Lowe S W 2003 Apoptosis and melanoma chemoresistance *Oncogene* **22** 3138–51

[23] Peach A H S 2006 Cutaneous melanoma *Surgery* **24** 21–6

[24] Ivanov V N, Bhoumik A and Ronai Z E 2003 Death receptors and melanoma resistance to apoptosis *Oncogene* **22** 3152–61

[25] Lin L *et al* 2023 Cell membrane oscillations under radiofrequency electromagnetic modulation *Langmuir* **39** 3320–31

[26] Yang X *et al* 2020 Cold atmospheric plasma induces GSDME-dependent pyroptotic signaling pathway via ROS generation in tumor cells *Cell Death Dis.* **11** 295

[27] Tanaka H, Mizuno M, Ishikawa K, Nakamura K, Kajiyama H, Kano H, Kikkawa F and Hori M 2011 Plasma-activated medium selectively kills glioblastoma brain tumor cells by down-regulating a survival signaling molecule, AKT kinase *Plasma Med.* **1** 265–77

[28] Yan D, Talbot A, Nourmohammadi N, Cheng X, Canady J, Sherman J and Keidar M 2015 Principles of using cold atmospheric plasma stimulated media for cancer treatment *Sci. Rep.* **5** 18339

[29] Cheng X, Sherman J, Murphy W, Ratovitski E, Canady J and Keidar M 2014 The effect of tuning cold plasma composition on glioblastoma cell viability *PLoS One* **9** e98652

[30] Yan D, Lin L, Xu W, Nourmohammadi N, Sherman J and Keidar M 2019 Cold plasma-based control of the activation of pancreatic adenocarcinoma cells *J. Phys. D: Appl. Phys.* **52** 445202

[31] Mohades S, Laroussi M, Sears J, Barekzi N and Razavi H 2015 Evaluation of the effects of a plasma activated medium on cancer cells *Phys. Plasmas* **22** 122001

[32] Lin L and Keidar M 2021 A map of control for cold atmospheric plasma jets: from physical mechanisms to optimizations *Appl. Phys. Rev.* **8** 011306

[33] Hanahan D and Weinberg R A 2011 Review 'Hallmarks of cancer: the next generation' *Cell* **144** 646–74

[34] Yao X, Lin L, Soni V, Gjika E, Sherman J H, Yan D and Keidar M 2020 Sensitization of glioblastoma cells to temozolomide by a helium gas discharge tube *Phys. Plasmas* **27** 114502

[35] Yao X, Yan D, Lin L, Sherman J H, Peters K B, Keir S T and Keidar M 2022 Cold plasma discharge tube enhances antitumoral efficacy of temozolomide *ACS Appl. Bio. Mater.* **5** 1610

[36] Soni V, Adhikari M, Simonyan H, Lin L, Sherman J H, Young C N and Keidar M 2021 In vitro and *in vivo* enhancement of temozolomide effect in human glioblastoma by non-invasive application of cold atmospheric plasma *Cancers (Basel)* **13** 4485

[37] Hattori N *et al* 2015 Effectiveness of plasma treatment on pancreatic cancer cells *Int. J. Oncol.* **47** 1655–62

[38] Adachi T, Tanaka H, Nonomura S, Hara H, Kondo S I and Hori M 2015 Plasma-activated medium induces A549 cell injury via a spiral apoptotic cascade involving the mitochondrial-nuclear network *Free Radic. Biol. Med.* **79** 28–44

[39] Yan D, Cui H, Zhu W, Talbot A, Zhang L G, Sherman J H and Keidar M 2017 The strong cell-based hydrogen peroxide generation triggered by cold atmospheric plasma *Sci. Rep.* **7** 10831

[40] Yan D, Lin L, Xu W, Nourmohammadi N and Jonathan H 2018 Universality of micro-molar-level cell-based hydrogen peroxide generation during direct cold atmospheric plasma treatment *Plasma Med.* **8** 335–43

[41] Nasri Z, Bruno G, Bekeschus S, Weltmann K D, von Woedtke T and Wende K 2021 Development of an electrochemical sensor for *in situ* monitoring of reactive species produced by cold physical plasma *Sens. Actuators* B **326** 129007

[42] Yan D, Xu W, Yao X, Lin L, Sherman J H and Keidar M 2018 The cell activation phenomena in the cold atmospheric plasma cancer treatment *Sci. Rep.* **8** 15418

[43] Yan D, Lin L, Xu W and Nourmohammadi N 2019 Cold plasma-based control of the activation of pancreatic adenocarcinoma cells *J. Phys. D: Appl. Phys.* **52** 445202

[44] Malyavko A, Yan D, Wang Q, Klein A L and Patel K C 2020 Cold atmospheric plasma cancer treatment, direct. Versus indirect approaches *Mater. Adv.* **1** 1494

[45] Yan D, Malyavko A, Wang Q, Ostrikov (Ken) K, Sherman J H and Keidar M 2021 Multimodal biological destruction by cold atmospheric plasma: capability and mechanism *Biomedicines* **9** 1259

[46] Privat-maldonado A, Connell D O, Welch E and Vann R 2016 Spatial dependence of DNA damage in bacteria due to low- temperature plasma application as assessed at the single cell level *Sci. Rep.* **6** 35646

[47] Ishaq M, Evans M D M and Ostrikov K K 2014 Atmospheric pressure gas plasma-induced colorectal cancer cell death is mediated by Nox2-ASK1 apoptosis pathways and oxidative stress is mitigated by Srx-Nrf2 anti-oxidant system *Biochim. Biophys. Acta, Mol. Cell. Res.* **1843** 2827–37

[48] Ishaq M, Kumar S, Varinli H, Han Z J, Rider A E, Evans M D M, Murphy A B and Ostrikov K 2014 Atmospheric gas plasma-induced ROS production activates TNF-ASK1 pathway for the induction of melanoma cancer cell apoptosis *Mol. Biol. Cell* **25** 1523–31

[49] Plewa J M, Yousfi M, Frongia C, Eichwald O, Ducommun B, Merbahi N and Lobjois V 2014 Low-temperature plasma-induced antiproliferative effects on multi-cellular tumor spheroids *New J. Phys.* **16** 043027

[50] Ishaq M, Bazaka K and Ostrikov K 2015 Intracellular effects of atmospheric-pressure plasmas on melanoma cancer cells *Phys. Plasmas* **22** 122003

[51] Yang H, Lu R, Xian Y, Gan L, Lu X and Yang X 2015 Effects of atmospheric pressure cold plasma on human hepatocarcinoma cell and its 5-fluorouracil resistant cell line *Phys. Plasmas* **22** 122006

[52] Keidar M, Walk R, Shashurin A, Srinivasan P, Sandler A, Dasgupta S, Ravi R, Guerrero-Preston R and Trink B 2011 Cold plasma selectivity and the possibility of a paradigm shift in cancer therapy *Br. J. Cancer* **105** 1295–301

[53] Vandamme M, Robert E, Dozias S, Sobilo J, Lerondel S, Le Pape A and Pouvesle J-M 2011 Response of human glioma U87 xenografted on mice to non-plasma treatment *Plasma Med.* **1** 27–43

[54] Lu X, Keidar M, Laroussi M, Choi E, Szili E J and Ostrikov K 2019 Transcutaneous plasma stress: from soft-matter models to living tissues *Mater. Sci. Eng. R. Rep.* **138** 36–59

Michael Keidar, Dayun Yan and Li Lin

Chapter 7

Selective anti-cancer performance

Under the same experimental conditions, a cold atmospheric plasma (CAP) treatment can selectively kill cancer cells over normal counterpart cells. Such a selective killing effect is a pivotal feature compared with many chemotherapeutic and radiotherapeutic modalities. Even if high dose CAP treatment causes damage to normal cell lines, it is possible to design a fractional treatment plan to avoid these side effects and target malignant cells. Understanding the selective mechanisms will build a foundation for CAP-based cancer therapy. In chapter 7, we provide forward-looking perspectives to re-define selectivity. Furthermore, we argue that the newly established physically-based treatment strategy provides an unprecedented vision to realize selectivity beyond the traditional concepts based on reactive species.

7.1 Selective anti-cancer effect

Cancer therapy is a main challenge in medicine. Many radiotherapeutic and chemotherapeutic modalities generate side effects or even significant harm to patients [1, 2]. We are in need of novel selective anti-cancer modalities as these have the most attractive clinical potential. CAP has demonstrated this potential over the past years, particularly considering plenty of *in vitro* studies performed on both traditional cell culture and 2D/3D cell culture using spheroids/scaffolds [3–7]. Currently, one of the most attractive features of CAP-based cancer therapy lies in its selective anti-cancer capacity that has been demonstrated in many cases. CAP's selectivity does not guarantee that normal cells will not be impacted. But with deep understanding of what drives preferential killing of cancer cells over normal cells, we can build methods that use fractional treatment minimize or eliminate unintended harm to normal cells. For example, 5 min of CAP treatment may cause a 90% and

doi:10.1088/978-0-7503-5537-7ch7

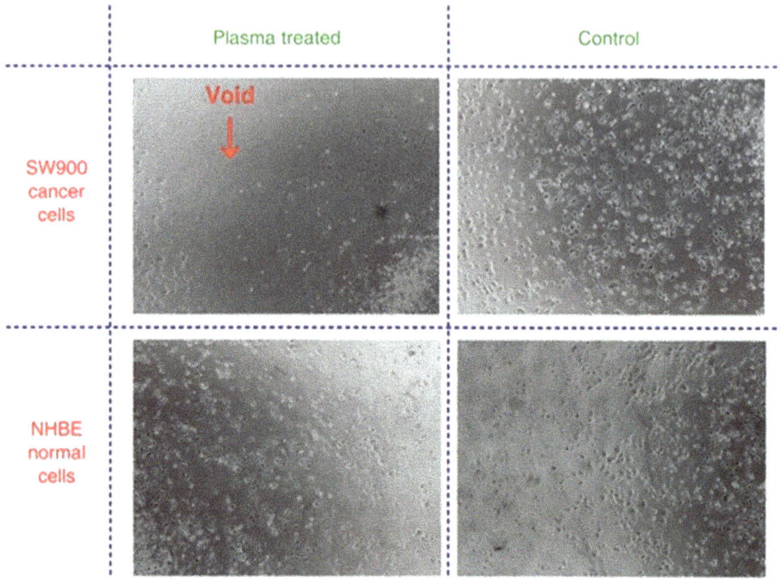

Figure 7.1. Selectivity effect of CAP treatment on SW900 cancer cells and normal NHBE cells *in vitro*. SW900 cells were detached from the plate in the CAP-treated zone, whereas no detachment was observed in the CAP-treated zone for NHBE cells. Reprinted from [12], copyright (2011), with permission from Springer Nature.

30% killing rate on cancer and normal cell line, respectively. Meanwhile, a one-minute CAP treatment may cause a 30% and 0% killing rate on cancer and normal cell lines, respectively. A good experimental design, such as five times fractional 1 min CAP treatment, may also achieve a 90% killing rate on cancer cells while leaving normal cells safe.

To date, the concept of selectivity is widely understood to mean that a CAP treatment, either direct or indirect, can cause cell death in cancer cell lines while only causing slight damage in normal cell lines, under the same experimental conditions [8]. One example is demonstrated in figure 7.1. In most cases, these *in vitro* tests were performed by treating a single cancer cell line or normal cell line under monoculture conditions. A selective anti-cancer effect was initially reported by Georgescu *et al*, Kim *et al*, and Zirnheld *et al* in their works about murine melanoma cells and human colon cancer cells in 2010 [9–11]. A summary of some typical selective cases was listed in table 7.1. The similarities among these widely observed features suggest some common biological mechanisms or responses of cancer cells and normal cells to CAP.

7.2 Selective anti-cancer mechanism

Selective death of cancer cells may be due to selective apoptosis after CAP treatment [4, 9, 10, 26]. Generally, cancer cells have a robust metabolism, so the basal ROS level is naturally higher in cancer cells than their counterparts [27–29]. When additional ROS stress is exerted, cancer cells tend to be killed more easily than

Table 7.1. Some selective anti-cancer capabilities of *in vitro* demonstrations [13] John Wiley & Sons. Copyright 2021 Wiley-VCH GmbH.

Cancer cell lines	Normal cell lines	References
Breast cancer (MCF-7)	Breast (M5655)	[14]
Benign prostatic hyperplasia (BPH-1)	Prostate primary epithelial cells	[15]
Colorectal cancer (HCT116)	Colon epithelial (FHC)	[16]
Cervical cancer (HeLa)	Fibroblast (HFB)	[17]
Gastric cancer (SC-2-NU)	Human fibroblast (WI-38)	[18]
Glioblastoma (U251SP)	Astrocytes (ACBRI-371)	[19]
Head and neck squamous carcinoma (JHU-022)	Oral cavity epithelial (OKF6)	[20]
Hepatoma cancer (HepG2)	Liver (L-02)	[21]
Lymphoma (U937)	Blood primary monocyte cells	[22]
Lung cancer (H460)	Lung fibroblast (L132)	[23]
Ovarian clear carcinoma (TOV21G)	Peritoneal mesothelial (OHFC)	[24]
Squamous cell carcinoma (PAM212)	Wild type keratinocytes (WTK)	[25]

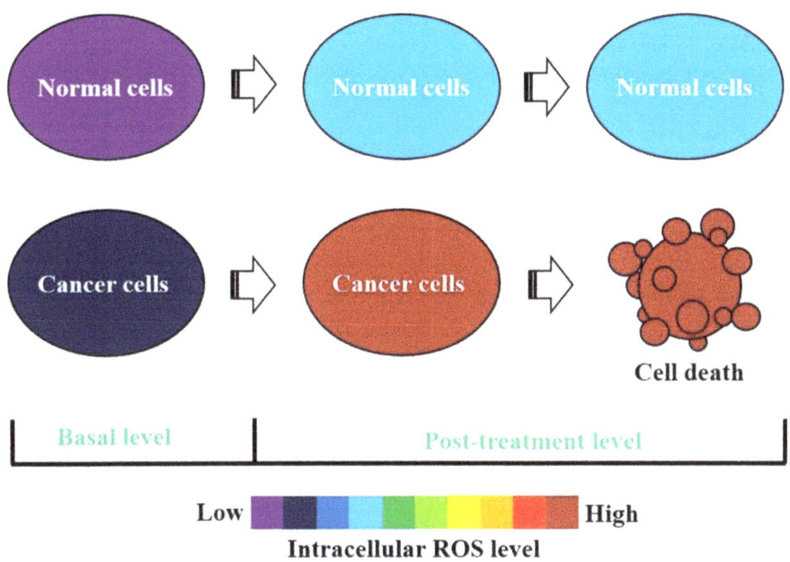

Figure 7.2. A schematic illustration of selective ROS rises in cancer cells after CAP treatment [13] John Wiley & Sons. Copyright 2021 Wiley-VCH GmbH.

normal cells because of the higher total ROS level (figure 7.2) [30, 31]. In practice, although the basal ROS level is difficult to determine accurately by common assays such as fluorescent assays [32]. The selective rise of intracellular ROS is the most widely observed cancer cellular change following CAP treatment [16, 32–35]. Selective intracellular ROS increase in the CAP-treated cancer cells may be the pivotal clue to understand CAP's selectivity [16, 32–35].

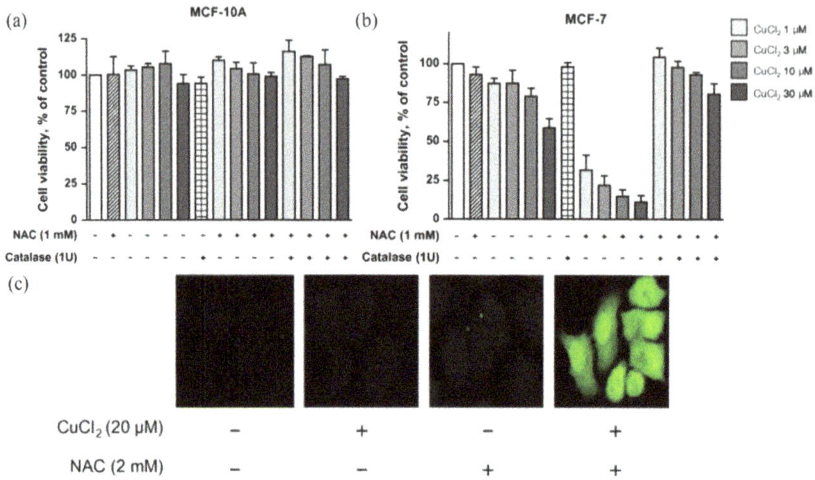

Figure 7.3. Selective anti-cancer effect based on H_2O_2 generated from NAC/Cu(II) [39]. (a) The neoplastic breast cancer line MCF-7. (b) The immortalized breast line MCF-10A. MCF-10A cells and MCF-7 cells were treated with 1 mM NAC and $CuCl_2$ (1–30 μM). Cell viability was determined by MTS assay. Data are expressed as percentages of the level detected in untreated control cells. (c) NAC/Cu(II) enhances intracellular ROS level in ovarian carcinoma cell line A2780. Cells were loaded with a carboxy-H2-DCFDA indicator (25 μM, ex/em: 495/529 nm) and treated with $CuCl_2$, NAC, or combination for 30 min. Reproduced from [39], copyright (2010), with permission from Elsevier.

H_2O_2 as a mediator of some anti-cancer drugs or alone at high dose can selectively induce apoptosis in some cancer cell lines [36–38]. For example, N-Acetylcysteine (NAC) reacts with copper (Cu) to generate H_2O_2 which has shown selective induction of cell death in human breast cancer cell line MCF-7 but not in normal breast cell line MCF-10A (figure 7.3). In plasma medicine, however, the observable selectivity cannot be chemically explained by H_2O_2 alone. More precisely, such a selective vision may not be explained or exactly predicted by just a single factor. First, CAP has a complex chemical and physical nature. In plasma medicine, most studies focus on the role of a few reactive species in selectivity, such as H_2O_2 and $^1O_2$, even though other ROS, RNS, and physical factors may also contribute to the final cellular responses to CAP. Second, cell populations are heterogeneous and even a single cell is an extremely complex biological system. Thousands of proteins or other molecules may have different expression patterns in cancer cells and normal cells, potentially contributing to selectivity.

We would like to emphasize that any models or explanations need to be consistent with the experimental trends and current findings in plasma medicine. More importantly, models of selectivity need to have universality, because of the consistent reports of selectivity in many cases. For example, selectivity also exists when the CAP-activated solutions were used to affect cancer cell lines [19, 40–46]. This suggests that the short-lived reactive species may not play a major role in the selectivity of PAS. In short, mechanisms to consider should account for features common in either cancer cells or normal cells.

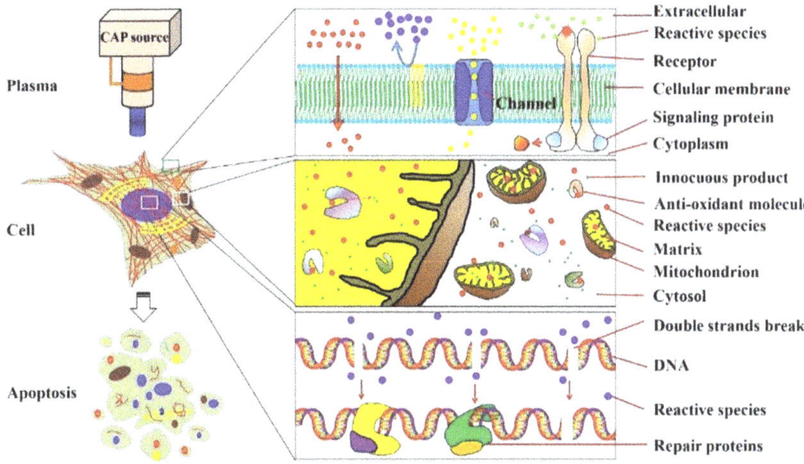

Figure 7.4. Selectivity may be affected by multiple factors and their specific expression levels in cancer cells and normal cells. The main factors may include reactive species diffusion, cellular membrane integrity, intracellular redox balance, and genome repairing systems. Reprinted from [8], with the permission of AIP Publishing.

Over the past years, several cellular factors have been proposed to explain selectivity, with most focused on the selective rise of intracellular ROS. Other specific cellular features may also provide clues to fully understand CAP's selectivity (figure 7.4). Despite lacking concrete experimental evidence for most models, current preliminary studies support some of these proposals.

7.3 H_2O_2 channels: aquaporins (AQP)

We realized the potential role of AQP in CAP-based cancer treatment. Recently, a selectivity model based on the differential expression of AQP in cancer and normal cells has been proposed [8]. AQP facilitates transmembrane diffusion of certain small molecules, including H_2O, CO_2, NO, NH_3 [47, 48]. New studies have confirmed that some AQP (1, 3, 5, 8, 9) also facilitated H_2O_2 passive membrane diffusion in bacteria, plant, and mammalian cells [49–53]. Furthermore, increased expression of AQP in cells enhanced H_2O_2 diffusion and intracellular ROS accumulation. For example, H_2O_2 uptake in yeast cells was drastically enhanced with additional AQP5 expressions (figure 7.5). In yeast cells, AQP8 expression caused a strong cytotoxicity in 0.2 mM H_2O_2 solution. In contrast, a 1.0 mM H_2O_2 solution could not cause noticeable death in yeast cells lacking AQP8 [52]. Of note, different AQP allow passage of different molecules at varying diffusion rates. AQP1 shows a much weaker H_2O_2 channel function compared to AQP8, which may be due to the smaller selectivity filter regions in AQP1. Together, these studies demonstrate that specific AQP's expression level can affect H_2O_2 cytotoxicity on cells [54].

Human tissues have specific AQP expression patterns. For example, AQP1, 8, and 9 are the main AQP expressed in the liver. AQP4 is the main AQP expressed in

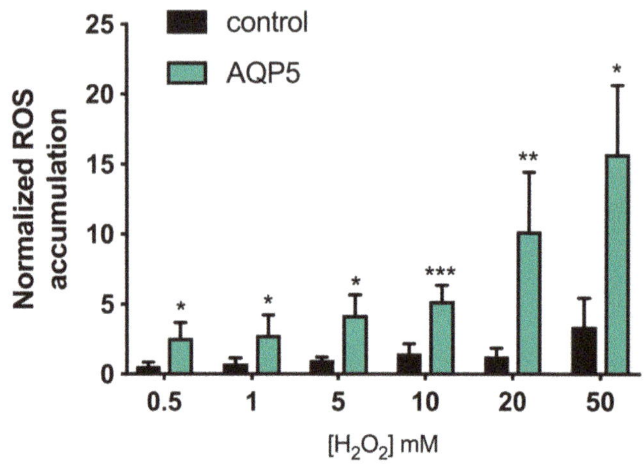

Figure 7.5. AQP5-facilitated H_2O_2 consumption of yeast cells is concentration-dependent. Yeast cells were incubated with H_2O_2 (0.5–50 mM). Intracellular ROS accumulation was recorded over 60 min, following H2-DCFDA fluorescence signal. Reproduced from [53]. CC BY 4.0.

the eyes, nose, and heart. Many mammalian tissues express AQP with H_2O_2 channel function, such as AQP1, or 3, or 8, including the eyes, brain, heart, lung, spleen, pancreas, liver, kidney, ovary, and muscle [55]. These tissues and cells are used in many CAP-based anti-cancer treatment studies. In cancer biology, AQP plays pivotal roles in tumor-associated edema, tumor proliferation, migration, and angiogenesis. Many tumorous tissues, such as lung cancer, breast cancer, astrocytoma, and cervical cancer, express more AQP1, 3, or 8 than their healthy tissues counterparts [56].

We hypothesized that the difference in ROS transmembrane diffusion rate between cancer cells and normal cells contributed to the selective rise of intracellular ROS in cancer post CAP treatment. To sustain cancer cells' vigorous metabolism and proliferation, many cancer cells express more AQP than normal cells. As a consequence, the faster uptake of H_2O_2 in cancer cells may contribute to the selective death in cancer cells. Though direct experimental evidence is still lacking in plasma medicine, some discussions can be given based on references and preliminary studies.

To date, the impact of AQP in CAP's cytotoxicity has only been explored in glioblastoma cell line U87MG. If we can inhibit AQP expression or block the diffusion of H_2O_2 through AQP, we can understand the impact of AQP on CAP's performance. Aquaporin blockers, such as Auphen, mercury (Hg), and silver (Ag), are widely used in the lab to block AQP [57]. Methods to inhibit AQP expression include gene silencing technology and genetic CRISPR-Cas9 knockout. An example is illustrated here. It has been found that AQP3 and AQP5 play a critical role in the H_2O_2 uptake in pancreatic cell line BxPC3 [53]. Either silencing the expression of AQP3/5 or both (siAQP3/siAQP5) will drastically weaken H_2O_2 flux into BxPC3 cells (figures 7.6(a) and (b)). Similarly, using AQP blockers, Auphen and Hg, can achieve an even bigger blocking effect, particularly when using $HgCl_2$ (figures 7.6(a) and (b)).

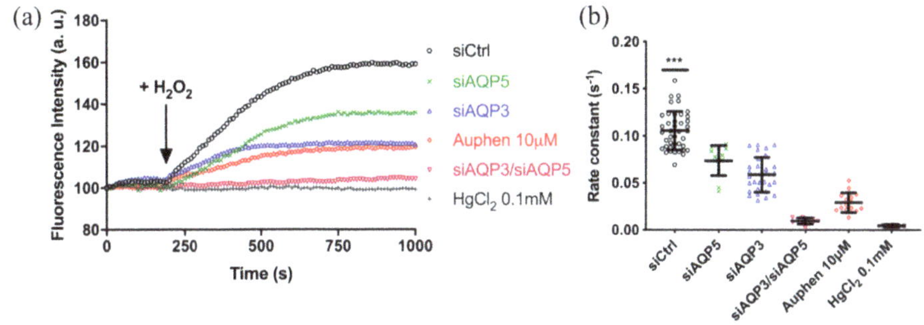

Figure 7.6. AQP3,5-mediated H_2O_2 uptake in pancreatic cancer cell line BxPC3 [53]. (a) Representative traces of H_2-DCFDA fluorescence after the addition of 100 μM extracellular H_2O_2. (b) First-order kinetic rate constant (s^{-1}) of H_2O_2 influx through endogenous AQP3 and AQP5 expressed in BxPC3 cells after addition of 100 μM extracellular H_2O_2. Experimental details were illustrated in reference. Reproduced from [53]. CC BY 4.0.

In plasma medicine, aquaporin blocker, Ag atom in $AgNO_3$, can inhibit CAP's cytotoxicity on U87MG cells [58]. U87MG cells have a clear expression of AQP1, 3, 8, and 9 [58]. Silencing the expression of AQP8 by transfecting cells with targeted miRNA significantly compromised CAP treatment efficacy [58], but silencing AQP9 did not change CAP's cytotoxicity on U87MG cells. These studies demonstrate that the specific role of AQP family channels in CAP's anti-cancer capacity is still far from clear. Several specific AQP may dominate the diffusion of reactive species and determine the cellular response to CAP. Further systematic and comparative studies of the AQP channels in specific cancer and normal cell lines are urgent missions.

7.4 Antioxidant system

Balance between the uptake of exogenous ROS and the resistance from intracellular antioxidant systems may determine net intracellular ROS level [50]. Cellular ROS-scavenging capability may be directly related to the expression level of molecular components in the intracellular antioxidant pathways. A complete intracellular antioxidant system includes small molecules, like glutathione and NAD(P)H, ROS-scavenging enzymes (superoxide dismutase (SOD), catalases (CAT), glutathione peroxidases (GSHPx)/reductases (GSHR), and peroxiredoxin (Prx)) [59]. These antioxidants have different functions and distribution within the cytoplasm and organelles. For example, it is known that peroxiredoxin is the main H_2O_2 scavenger in mitochondria [60].

Given this rationale, a robust intracellular antioxidant system will protect cells from CAP's cytotoxicity. If a cell has a more robust antioxidant system, it can better resist the exogenous ROS attack, minimizing damage from low and intermediate dose CAP treatments. In one study, impairing expression of Cu, Mn, and Zn-SOD improved the death rate of CAP-treated HeLa cells [43]. The authors also found that catalase overexpression decreased the death rate of HeLa cells [43]. Several studies have found that the expression level of antioxidant enzymes is lower in many cancer cell lines compared to their normal counterpart cell lines [27]. For example, the

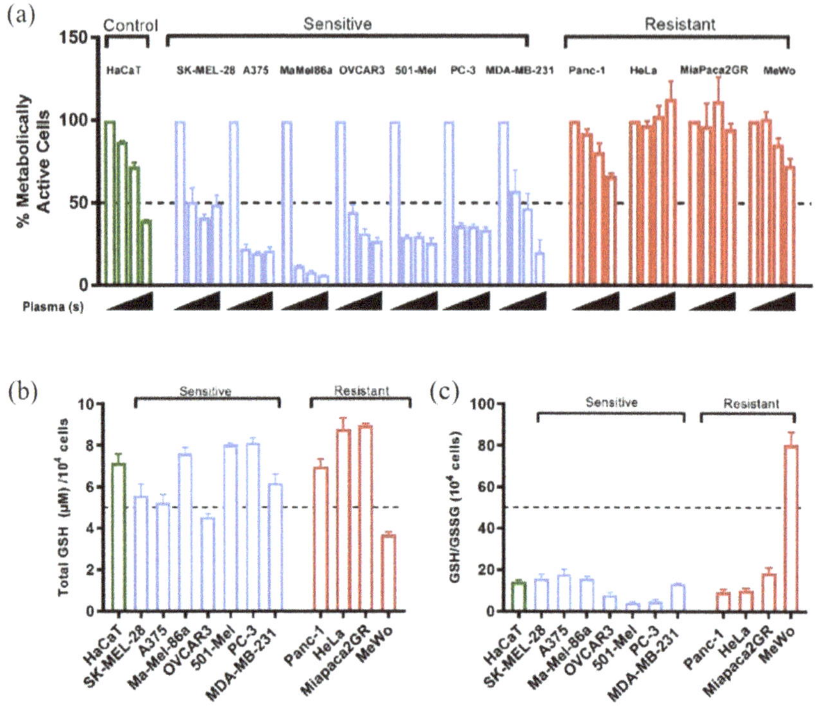

Figure 7.7. Specific sensitivity of cancer cell lines to CAP treatment [64]. (a) Metabolic activity at 24 h after CAP treatment (0 s, 30 s, 60 s, 120 s). Cell lines that showed >50% reduction in metabolic activity after CAP treatment (30 s) were categorized as 'sensitive', and <50% reduction were categorized as 'resistant' cell lines. (b) Basal glutathione (GSH) levels. (c) GSH/GSSG ratio. Reproduced from [64]. CC BY 4.0.

expression level of catalase in cancer cells is less than normal cells in many cases [61–63]. Another recent study found that stable intracellular glutathione levels after CAP treatment could only be observed in some ROS resistant cell lines. Compared with other ROS sensitive cell lines, these resistant cell lines have higher expression of cysteine-glutamate antiporter xCT, higher average glutathione (GSH) and average GSH/GSSG ratio (figure 7.7) [64].

Furthermore, there is an inverse correlation between a cancer cells' H_2O_2 consumption rate and CAP's cytotoxicity. Considering ten cancer cell lines, cancer cell lines with faster H_2O_2 consumption rates are more resistant to the same CAP treatment. For example, bone osteosarcoma cell line U-2 OS shows the slowest extracellular H_2O_2 scavenging rate and has shown the strongest vulnerability to CAP treatment. In contrast, melanoma cell line B16F10 can clear all extracellular H_2O_2 in under an hour after CAP treatment, the quickest among all ten cell lines tested here. Correspondingly, B16F10 cells show extreme resistance to CAP treatment (figure 7.8(a)). Such a quasi-inversely proportional correlation suggests a pivotal role of the intracellular antioxidant system in CAP's cytotoxicity.

Another general trend found is that cells expressing tumor suppressor *p53* are more resistant to CAP treatment than *p53*-deficient cells or cells with comprised

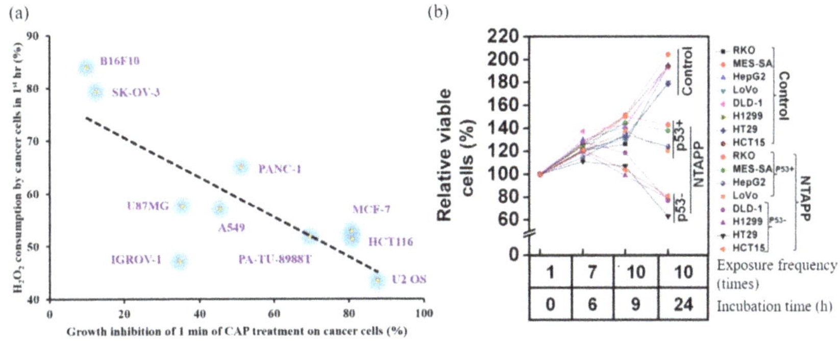

Figure 7.8. Two general trends observed *in vitro*. (a) Extracellular H_2O_2 consumption rate of a cancer cell line is inversely proportional to killing rate. Copyright 2018 IEEE. Reprinted, with permission, from [65]. (b) Preferential anti-cancer efficacy on cancer cells without functional p53. Here, RKO, MESSA, HepG2, G361, LoVo are p53-proficient cancer cell lines. DLD-1, H1299, HT29, HCT115 are p53-deficient cancer cell lines. NTAPP: nonthermal atmospheric pressure plasma. Reproduced from [66]. CC BY 4.0.

expression of *p53* (figure 7.8(b)) [66]. *p53* is a critical transcription factor with multiple functions and loss of *p53* is a pivotal step in many tumors' tumorigenesis. It is common to find comprised *p53* expression in the growth stage of many aggressive tumors [67]. One of its important functions related to CAP is regulating the expression of a series of antioxidant enzymes that further protects the genome from ROS-mediated oxidation [68–71]. Because *p53* is an important regulator of intracellular ROS levels [72], cancer cells lacking *p53* or cells that have compromised expression levels will have a weaker antioxidant system, potentially explaining the stronger cell death rate in *p53*-deficient cells after CAP treatment.

7.5 Transient membrane pores

Transient pore formation in cellular membranes, particularly the cytoplasm membrane, provides another perspective to understand selectivity. Based on the kinetics of transient membrane pore formation on the cytoplasmic membrane, molecular dynamics (MD) simulations can provide new clues. It has been found that the oxidation of lipids in the phospholipid bilayer causes a drop in the electric field threshold, a physical property that also plays a role in membrane pore formation. The newly formed pores are able to facilitate the penetration of extracellular ROS [73]. In addition to considering the electric field, MD simulation reveals that a transient membrane pore can be formed once all phospholipids have been oxidized [74]. And we know that CAP-originated ROS can oxidize phospholipids of the cytoplasmic membrane.

More importantly, MD simulation studies suggest a unique role of cholesterol in the movement of reactive species across the phospholipid bilayer. A higher cholesterol fraction in the phospholipid bilayer increases its order and free energy transfer across barrier height or width. This results in a local free energy minimum in the membrane's center to create extra free energy barriers (figure 7.9) [74, 75].

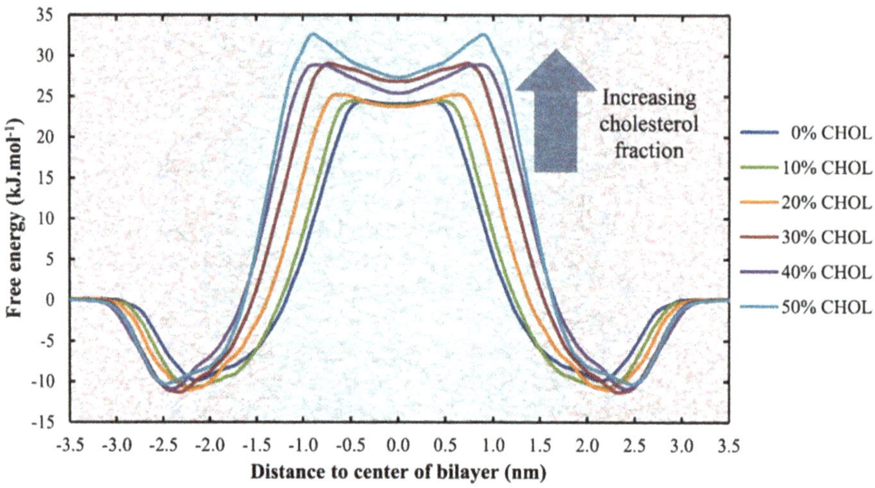

Figure 7.9. Effect of cholesterol fraction in cell membrane on the potential of mean force (PMF) of H_2O_2. Reproduced from [75]. CC BY 4.0.

Therefore, cholesterol may protect the cytoplasmic membrane from transient pore formation due to lipid peroxidation [74, 75]. Interestingly, cholesterol may have a specific distribution whether in cancer cells or normal cells. Compared with normal lymphocytes, a lower cholesterol/phospholipid ratio in leukemic cells has been observed [76]. Given these considerations, the CAP-originated reactive species may diffuse into cancer cells more easily than normal cells due to formation of transient pores in cytoplasmic membrane. However, more references are needed to fully support the assumption that the cholesterol/phospholipid ratio in cancer cells is lower than normal cells. And, currently, the evidence is limited to lymphocytes and lymphoma cell types [77].

7.6 Singlet oxygen

In addition to H_2O_2, CAP generates other ROS, such as singlet oxygen ($^1O_2$), which may have a significant impact on cells [78, 79]. $^1O_2$ inactivates membrane-bound catalase (CAT), generates cancer cell-derived secondary $^1O_2$, and initiates ROS/RNS-dependent apoptotic events [80]. Unlike other reactive species, $^1O_2$ pathways may result in self-perpetuating apoptotic signaling from cell to cell potentially reaching deep tumor tissue, which cannot be affected by a remote CAP treatment [81]. $^1O_2$ from CAP may initially only reach a tumor's surface and inactivate some membrane-associated CAT. After that, at the local CAT inactivation site, secondary $^1O_2$ may be generated by cancer cells, which may further inactivate other CAT on the same cells and neighboring cells. The generation of secondary $^1O_2$ and CAT inactivation may continue to spread into deeper tissues. Finally, apoptotic cells appear in the upper layers of deeper tissue. Given specific expression patterns of membrane-bound catalase, this singlet oxygen model can explain the selectivity on cancer cells *in vitro* [81].

To support this model, evidence must demonstrate that membrane-bound catalase expression is specific, in some way, on cancer cells versus normal cells. Another challenge will be to elucidate the unclear role of $^1O_2$ in CAP cancer treatment. In other words, stronger evidence is needed to demonstrate that $^1O_2$ indeed plays a pivotal role, at least as important as H_2O_2, in determining the final fate of CAP-treated cancer cells at least *in vitro* [22, 23, 82–88]. Furthermore, as a short-lived reactive species, $^1O_2$ has a half-life of just several microseconds in water [89]. So, it is not expected to play a role in the selectivity of either CAP-activated solution (PAS) or medium (PAM), which theoretically only retains long-lived reactive species or other reaction products to affect cancer cells [19, 40–46]. In short, more investigation of $^1O_2$'s unique roles in direct CAP treatment is warranted.

7.7 Genomic repairing system

DNA damage, an early event observed in many treatments, is an important consequence that can trigger apoptosis in CAP-treated cancer cells [90, 91]. Among the different types of DNA damage, double-strand break (DSB) is a main DNA damage seen in plasma medicine [92, 93]. One well-studied DSB marker is the phosphorylation of serine 139 on H_2AX histone (*γ-H2AX*), which has been widely observed immediately following CAP treatment [92–94]. Increased expression of phosphorylated *p53* (*p-p53*) in the CAP-treated melanoma cells (B16F10) was followed by the *γ-H2AX* marker. These observations, align with the canonical order of DNA damage-triggered apoptosis pathways [95]. If DNA damage can be repaired, however, apoptosis may not be triggered. In mammalian cells, high-fidelity genome repairing systems can resist DNA damage, to some extent [96]. For normal cells, inefficient DNA damage surveillance and repair will lead to the accumulation of mutations, and if sufficient, may initiate either tumorigenesis or apoptosis [97, 98]. Like normal cells, severe DNA damage in cancer cells will also trigger apoptotic pathways. But in cancer cells, ineffective DNA damage repair will accumulate mutations that may facilitate tumor growth and aggression [99–101]. In other words, the genome repairing system has been compromised in many tumors, which in turn may facilitate the increased frequency of selective DNA damage and apoptosis in the CAP-treated cancer cells [102]. For example, isothiocyanates' (ITCs) selective anti-cancer capability was due to differing ability of prostate cancer cells (PC-3) and normal dermal fibroblasts (HDFa) to handel DNA damage (figure 7.10) [103].

Here, we need to emphasize that selectivity does not occur in all cases because different cells have different expression levels of relevant molecular features. The expression patterns of specific proteins or protein families are drastically different between cancer and normal counterparts. The previously explained AQP is a prime example. If specific cancer cells express H_2O_2 channels AQP1, 3, 5, 8, or 9 less than normal cells, then CAP may kill cancer cells with a negative selectivity. Similarly, for the explanation based on the transient pore formation mechanism, if the specific cancer cells have more cholesterol in their cellular membrane than normal counterparts, then a negative selectivity may also be observed.

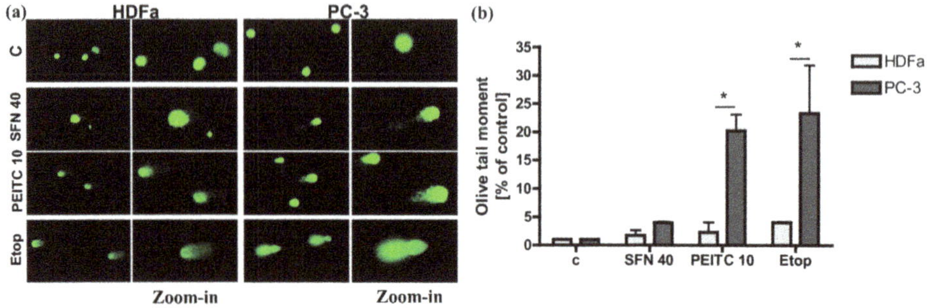

Figure 7.10. Selective anti-cancer activity of isothiocyanates (ITCs) on prostate cancer cells (PC-3) and normal dermal fibroblasts (HDFa) is due to different DNA damage repairing between cancer and normal cells [103]. (a) DNA damage was diagnosed by Alkaline comet assays. Three ITCs (SFN (40 μM), etoposide (20 μM), PEITC (10 μM)) treated cells for 3 h. In control group, cells were treated by DMSO. (b) Olive tail moment was calculated to assess DNA integrity. Reproduced from [103].

7.8 Novel selectivity concepts

Common definition of selectivity is based on a single cell line's response to reactive species or other CAP-specific chemical modification in solutions. Here we identify several novel approaches to extend the definition of selectivity, particularly based on recent studies.

7.8.1 Co-culture of cancer and normal cells *in vitro*

Considering its potential applications, selective cancer cell death needs to be fully characterized, especially at the interface between cancer tissues and normal tissues. Co-culture of cancer and normal cells *in vitro* can simulate the interface and environment found *in vivo*. One can hypothesize that if cancer cells can consume ROS/RNS faster than normal cells, they may actually protect normal cells in a co-culture system. For example, we observed glioblastoma cell line U87MG consumed H_2O_2 much faster than normal astrocyte cell line hTERT/E6/E7 [8]. This hypothesis can explain the selective anti-cancer performance of a co-culture system composed of liver cancer cells HepG2 and normal cell line L0 [45]. In that study, an optimum CAP treatment dose selectively trigger significant apoptosis in cancer cells with limited damage to normal cells [45].

What's more, CAP's selectivity in a co-culture system can be achieved via self-adaptive guiding of the plasma plume away from normal cells, by using a specific physical measurement and feedback mechanism. In a recent new co-culture design (figure 7.11), glioblastoma cell line U87MG could direct a helium CAP plume either toward or away from normal astrocyte cell line hTERT/E6/E7 [104]. It is necessary to point out that the presence of a grounded copper sheet beneath the cell culture dish could change CAP plume's behaviors compared with a floating condition without a copper sheet [104]. We proposed that cellular capacitance may be an essential factor affecting the CAP plume's behaviors [104]. This example was the earliest demonstration that a CAP plume's direction could be manipulated by a

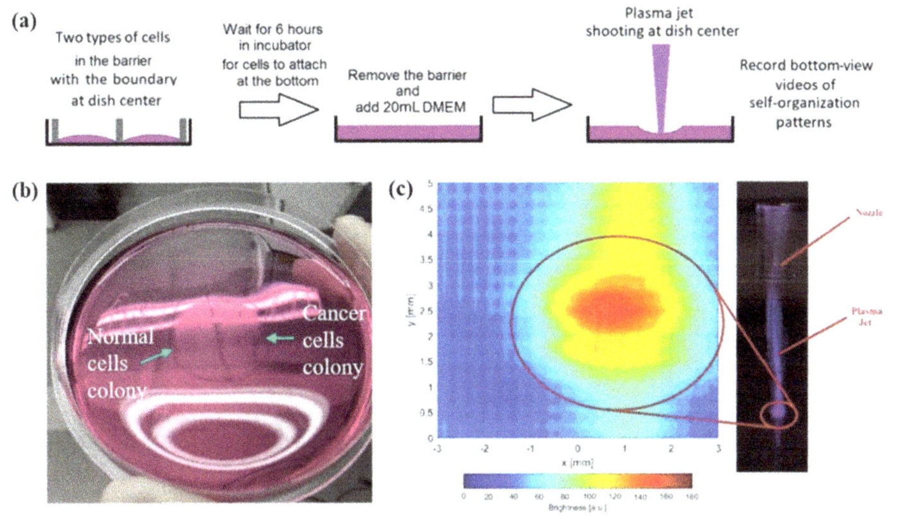

Figure 7.11. Plasma plume was directed to the boundary of cancer cell colonies and normal cell colonies in a co-culture system [104]. (a) Schematic illustration of co-culture system and experimental designs. (b) Photo of two cell colonies in 100 mm culture dish. (c) A side-view image of the plasma plume' contact area on targets. Normal astrocytes hTERT/E6/E7 and glioblastoma cells U87MG have grown on the negative-x (left) region and the positive-x (right) region, respectively. Reprinted with permission from [104], copyright (2019) American Chemical Society.

unique physical factor: the permittivity difference between cancer cells and normal cells. This pioneering work contributes to the foundation of an advanced understanding of CAP selectively to cancer tissue in the vicinity of normal tissues, something significant in clinical applications, particularly surgical cases.

7.8.2 Specific cell-based H_2O_2 generation

Recent observations provide another opportunity to extend the concept of selectivity. In previous studies, CAP has been regarded as the sole source of ROS affecting cellular function and final fate, particularly H_2O_2. Strong cell-based H_2O_2 generation at micromolar levels challenged this consensus. H_2O_2 is a critical intracellular signaling molecule in cancer cells [36]. As early as two decades ago, melanoma, neuroblastoma, ovarian carcinoma, and colon carcinoma were found generating nanomolar (nM) levels of H_2O_2 [105]. Prior to this, strong H_2O_2 generation from cancer cells has rarely been reported. To date, CAP may be the sole candidate able to trigger cancer cells to generate H_2O_2 generation at micromolar (μM) level [106]. This discovery is unique in plasma medicine and may provide inspiration for further investigations in cancer biology and therapy, because microlevel H_2O_2 is toxic to cancer cells.

Cell-based H_2O_2 generation was initially observed in pancreatic adenocarcinoma cell line PA-TU-8988T and triple-negative breast cancer cell line MDA-MB-231 by comparing H_2O_2 concentration in CAP-treated medium with and without cells [106]. When culture medium's volume was adequately small, cell-based H_2O_2 generation could reach about 60 μM (figure 7.12(a)). APPJ's

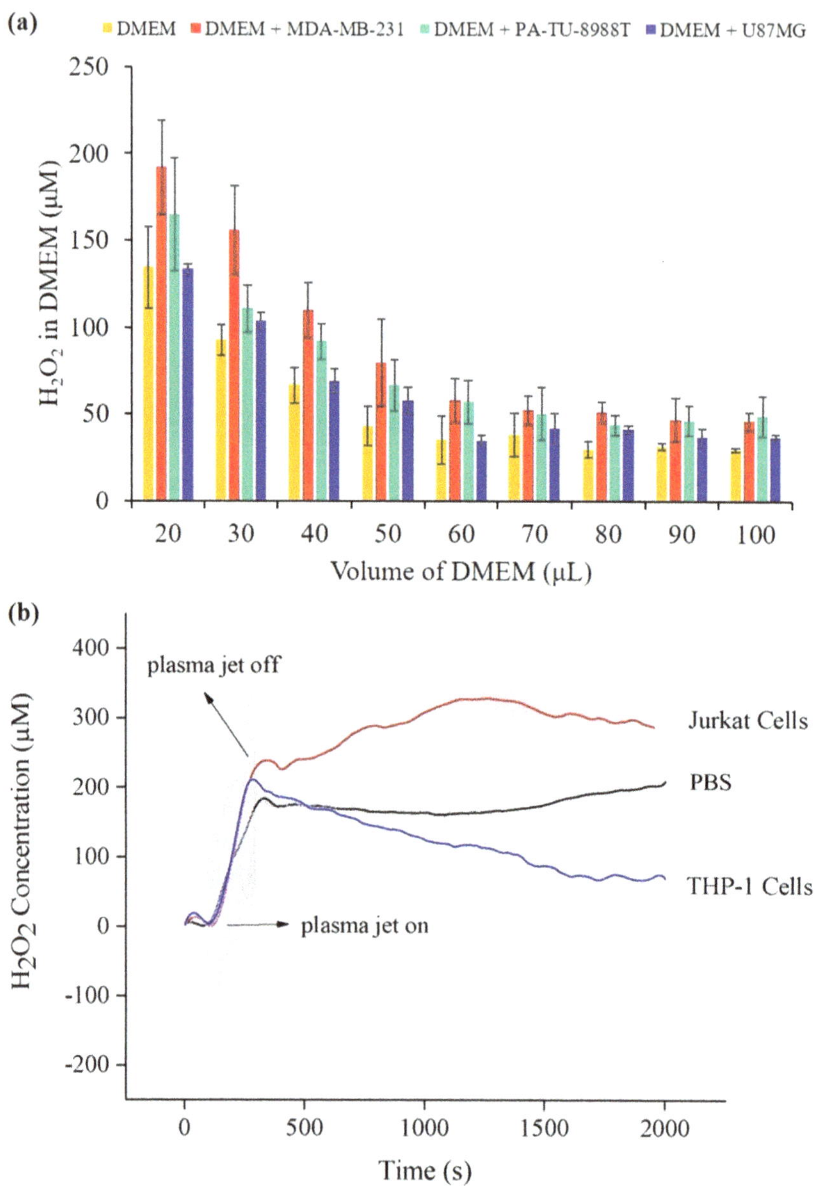

Figure 7.12. Specific strong cell-based H_2O_2 generation. (a) Direct treatment can specifically trigger H_2O_2 generation in the extracellular environment. H_2O_2 was measured immediately (<1 min) after treatment. Each trial with a specific volume of DMEM was performed based on following designs: just DMEM (control), pancreatic adenocarcinoma cells (PA-TU-8988T) in DMEM, breast adenocarcinoma cells (MDA-MB-231) in DMEM, and glioblastoma cells (U87MG) in DMEM. Reproduced from [106]. CC BY 4.0. (b) An *in situ* H_2O_2 measurement over 3 min Ar APPJ treatment and 30 min after the treatment. Jurkat T cells and THP-1 cells were human monocytic cell lines, and the cancer cell lines were derived from an acute monocytic leukemia patient, respectively. PBS was used as the control (pH 7.4). Reprinted from [108], copyright (2021), with permission from Elsevier.

discharge voltage is a critical parameter to modulate H_2O_2 generation. When the discharge voltage was high enough, noticeable cell-based H_2O_2 generation was observed in six cancer cell lines, including cervical adenocarcinoma (HeLa), bone osteosarcoma (U-2 OS), lung carcinoma (A549), colorectal carcinoma (HCT116), melanoma (B16F10), and breast cancer (MCF-7) [107]. In other recent studies, similar phenomena were observed by another team using *in situ* H_2O_2 measurement (figure 7.12(b)) [108].

Cell-based H_2O_2 generation may be cell line specific and contribute to selectivity for cancer cells over normal cells. Cancer cell lines have quite different cell-based H_2O_2 generation than normal cells, when tested under the same experimental conditions [108]. Compared to cancer cell lines, two normal cell lines, fibroblast cell line (WDTF) and astrocyte cell line (hTERT/E6/E7), generate only low concentration of H_2O_2 (<5 μM) under certain plasma discharge conditions [109]. A recent *in situ* study found that both an immortalized human CD4+ T lymphocyte cell line (Jurkat T cells) and a human monocytic cell line derived from an acute monocytic leukemia patient (THP-1 cells) generated H_2O_2. And although THP-1 cells generate H_2O_2 just during CAP treatment, Jurkat T cells generate H_2O_2 continuously after a CAP treatment. THP-1 cells did not generate H_2O_2 post treatment and therefore quickly consumed extracellular H_2O_2 (figure 7.12(b)). The similar fast H_2O_2 consumption was also observed in our studies on MBA-MD-231 and PA-TU-8988T cells [106]. Clearly, a selective cell-based H_2O_2 generation can be used to enhance the selectivity of direct CAP treatment.

7.8.3 Physically-based selectivity

In plasma medicine, the biological effects of CAP's physical factors are rarely reported. Particularly for cancer treatment, chemical factors are widely regarded to determine cellular fates both *in vitro* and *in vivo*. However, as previously discussed, physical factors in CAP can also contribute to selectivity by novel mechanisms. Physical factors mainly exert three effects on a biological system: ultravoiet (UV) effect, thermal effect, and general electromagnetic (EM) effects. Recent observations overturn previous assumptions that physical factors play negligible roles in CAP's biological effect. We systematically illustrate this topic in chapter 6. Two key results validated a significant physical mechanim of action by CAP: strong cell death due to trans barrier CAP treatment and sensitization of cancer cells to drugs after physically-based CAP treatment. Obviously, these observations provide an unprecedented vision to understand a much wider interaction between CAP and cells.

To isolate the physical effect, experiments must be designed to block all chemical factors. EM emission from APPJ penetrates the bottom of standard cell culture dishes and multi-well plates and can cause a strong physically-triggered cell death. Dishes and plates were made of polystyrene material with a thickness of ~1 mm. A simple experimental design change, inversion of cell culture dishes, blocked all chemical factors from contacting the cells on the other side (figure 7.13(a)). The main cell death triggered by physical factors is necrosis, characterized by apparent

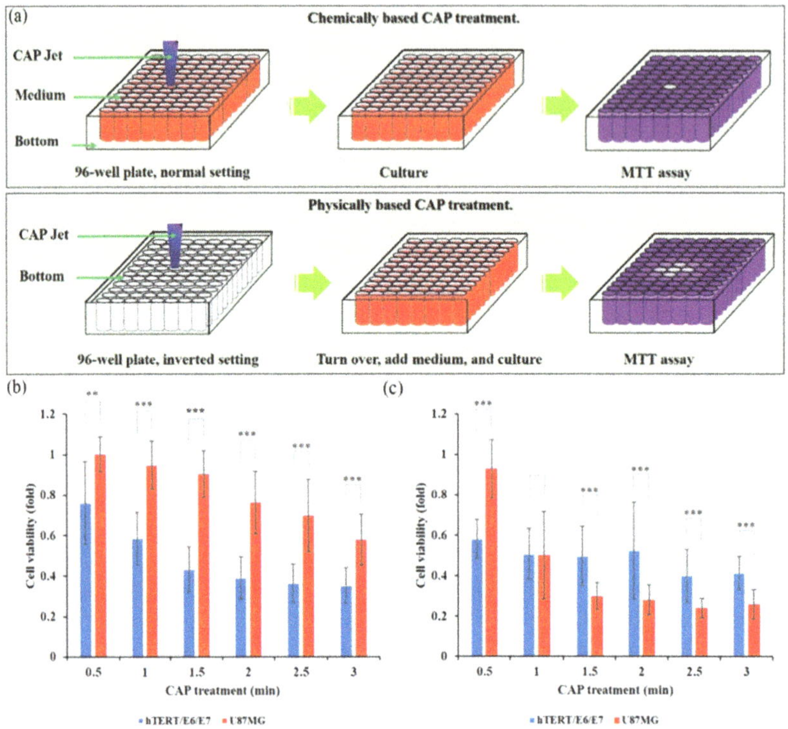

Figure 7.13. Physically-based treatment can reinstate the selectivity in the treated glioblastoma cells and astrocytes. (a) Schematic illustration of chemically- and physically-based treatment. (b) Negative selectivity in chemically-based treatment. (c) Positive selectivity in physically-based treatment [112].

cytosol aggregation and bubbling on the cytoplasmic membrane [110, 111]. The novel responses observed after physically-based treatment supports that separate pathways and mechanisms are triggered depending on the chemical and physical factors present. In one case, normal human astrocyte cell line hTERT/E6/E7 was actually more sensitive to reactive species' cytotoxicity than cancer cells. Negative selectivity was even observed, with human glioblastoma cell line U87MG surviving better than the normal cells (figure 7.13(b)) [112]. This limitation cannot be overcome with a solely chemically-based CAP approach. In contrast, physically-based CAP treatment reinstated a positive selectivity toward glioblastoma cells (figure 7.13(c)) [112]. Glioblastoma multiforme (GBM) is characterized as a highly invasive and aggressive brain tumor [113]. Individuals with GBM face a poor prognosis, with few surviving more than two years [113, 114]. Currently, a combination of different therapeutic modalities is the gold standard for GBM treatment, which typically includes chemotherapy, radiotherapy, and surgical resection. However, all these approaches have some drawbacks [113, 115, 116]. The significant penetration of physically-based treatment provides a novel non-invasive option to improve the therapeutic outcome of existing GBM treatment approaches, with high selectivity for cancer cells.

To date, almost all research strategies focused on killing cancer cells directly with CAP. We note that the relatively short effective peneration range of traditional CAP treatment is a big caveat in current plasma medicine. Recently, our work has shed light on a novel way to use CAP in cancer treatment and other potential medical fields. Not aiming to directly kill cancer cells, the discharge tube focuses on sensitizing cancer cells to certain drugs in a contactless and trans barrier way, providing a promising method to affect deep subsurface tumor tissues, such as brain tumor.

To illustrate, the helium was discharged in a sealed glass tube under atmospheric pressure conditions. We called this the discharge tube. Compared with traditional APPJ sources, the discharge tube was essentially the same except the helium was sealed in the tube rather than continuouesly supplied from a pressurized helium tank (figures 7.14(a) and (b)) [117]. Two glioblastoma cell lines U87MG and A172 were effectively sensitized to a widely used glioblastoma drug temozolomide (TMZ) (figure 7.14(c)). Interestingly, there was no apparent sensitization effect on normal astrocyte cell line hTERT/E6/E7 (figure 7.14(d)). Clearly, EM emission from the discharge tube triggered the sensitization process, though the complete picture of the physical basis and mechanism are still unknownn [117]. The sensitization could even be achieved through a trans barrier way, consisting of a macroscale air gap (5 mm)

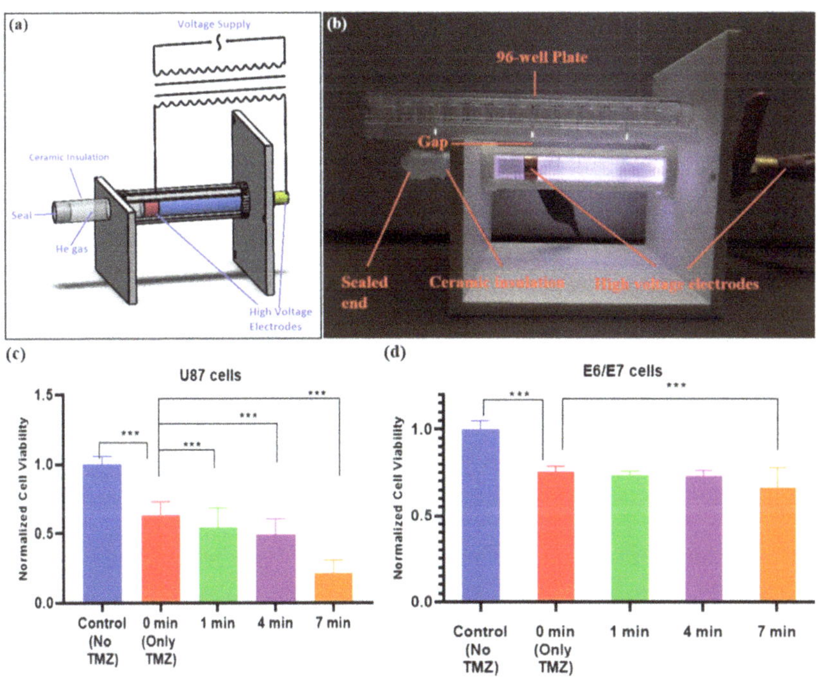

Figure 7.14. Selectively sensitizing glioblastoma cells using a helium gas discharge tube [117]. (a) Schematic illustration of a discharge tube. (b) Discharge tube in operation. (c) Strong sensitization of glioblastoma U87MG cells to TMZ (250 μM). (d) Weak sensitization of normal astrocyte cell line hTERT/E6/E7. Reprinted from [117], with the permission of AIP Publishing.

and a physical barrier with a thickness of 1 mm. In some references, sensitization is also referred to as activation and can be achieved by typical APPJ sources. The activation phenomena was first discovered during a direct CAP treatment using a helium APPJ. In one example, pancreatic cancer cells were sensitized to the cytotoxicity of long-lived reactive species H_2O_2 and NO_2^- [118, 119]. Although these activation mechanisms are also unknown, the short-lived reactive species and physical factors rather than long-lived reactive species should trigger the activation in a direct treatment.

7.9 Summary

Traditional definitions of CAP-caused death of cancer cells employed a restricted concept of selectivity. It focused only on the death rate of a single CAP treatment on a single cell line, which was fully determined by CAP's chemical effect. Today, plasma medicine has proposed several theories to explain selectivity. They involve, specific cellular targets, processes, or functions, such as: reactive species transmembrane diffusion, transition membrane pore formation, the intracellular antioxidant system, as well as DNA damage repair systems. In short, the selective anti-cancer capacity of CAP cannot be regarded as one-dimensional. Multiple factors, both chemical and physical, contribute to CAP's selective capacity.

Here, we expanded the traditional definition of selectivity and provided some new visions for the concept. Firstly, in co-culture systems, cancer cells may actually protect normal cells by consuming reactive species faster than normal cells. Second, the capacitance-based physical feedback mechanism connects information from cells to CAP sources and provides a routine to guide bulk plasma to target cancer cells in the presence of neighboring normal cells. Third, the new investigations on the anti-cancer physical effects, either direct killing or sensitization, provide new avenues to understand selectivity as well. Physically-based treatment generated positive selectivity in glioblastoma cells, a fact that could not be achieved with a chemically-based treatment. Finally, the discharge tube, a novel CAP source, sensitized glioblastoma cells to the efficacy of the widely used drug TMZ without causing side effects on normal astrocytes. These new studies present a clear vision of the CAP-based selective anti-cancer effect that goes beyond its previously limited definition and understanding.

References

[1] Meirow D and Nugent D 2001 The effects of radiotherapy and chemotherapy on female reproduction *Hum. Reprod. Update* **7** 535–43

[2] Coates A, Abraham S and Kaye S B *et al* 1983 On the receiving end—patient perception of the side effects of cancer chemotherapy *Eur. J. Cancer Clin. Oncol.* **19** 203–8

[3] Schlegel J, Köritzer J and Boxhammer V 2013 Plasma in cancer treatment *Clin. Plasma Med.* **1** 2–7

[4] Yan D, Sherman J H and Keidar M 2017 Cold atmospheric plasma, a novel promising anti-cancer treatment modality *Oncotarget* **8** 15977

[5] Keidar M 2015 Plasma for cancer treatment *Plasma Sources Sci. Technol.* **24** 033001

[6] Graves D B 2014 Reactive species from cold atmospheric plasma: implications for cancer therapy *Plasma Processes Polym.* **11** 1120–7

[7] Chauvin J, Gibot L, Griseti E, Golzio M, Rols M P, Merbahi N and Vicendo P 2019 Elucidation of *in vitro* cellular steps induced by antitumor treatment with plasma-activated medium *Sci. Rep.* **9** 4866

[8] Yan D, Talbot A, Nourmohammadi N, Sherman J H, Cheng X and Keidar M 2015 Toward understanding the selective anticancer capacity of cold atmospheric plasma–a model based on aquaporins *Biointerphases* **10** 040801

[9] Georgescu N and Lupu A R 2010 Tumoral and normal cells treatment with high-voltage pulsed cold atmospheric plasma jet *IEEE Trans. Plasma Sci.* **38** 1949–55

[10] Kim J Y, Kim S O, Wei Y and Li J 2010 A flexible cold microplasma jet using biocompatible dielectric tubes for cancer therapy *Appl. Phys. Lett.* **96** 203701

[11] Zirnheld J L, Zucker S N, DiSanto T M, Berezney R and Etemadi K 2010 Nonthermal plasma needle: development and targeting of melanoma cells *IEEE Trans. Plasma Sci.* **38** 948–52

[12] Keidar M, Walk R, Shashurin A, Srinivasan P, Sandler A, Dasgupta S, Ravi R, Guerrero-Preston R and Trink B 2011 Cold plasma selectivity and the possibility of a paradigm shift in cancer therapy *Br. J. Cancer* **105** 1295–301

[13] Yan D, Horkowitz A, Wang Q and Keidar M 2021 On the selective killing of cold atmospheric plasma cancer treatment: status and beyond *Plasma Processes Polym.* **18** 2100020

[14] Mirpour S, Ghomi H, Piroozmand S and Nikkhah M 2014 The selective characterization of nonthermal atmospheric pressure plasma jet on treatment of human breast cancer and normal cells *IEEE Trans. Plasma Sci.* **42** 315–22

[15] Hirst A M, Simms M S, Mann V M, Maitland N J, O'Connell D and Frame F M 2015 Low-temperature plasma treatment induces DNA damage leading to necrotic cell death in primary prostate epithelial cells *Br. J. Cancer* **112** 1536–45

[16] Ishaq M, Evans M D M and Ostrikov K K 2014 Atmospheric pressure gas plasma-induced colorectal cancer cell death is mediated by Nox2-ASK1 apoptosis pathways and oxidative stress is mitigated by Srx-Nrf2 anti-oxidant system *Biochim. Biophys. Acta, Mol. Cell. Res.* **1843** 2827–37

[17] Kwon B-S, Choi E H, Chang B, Choi J-H, Kim K S and Park H-K 2016 Selective cytotoxic effect of non-thermal micro-DBD plasma *Phys. Biol.* **13** 056001

[18] Torii K *et al* 2015 Effectiveness of plasma treatment on gastric cancer cells *Gastric Cancer* **18** 635–43

[19] Tanaka H, Mizuno M, Ishikawa K, Nakamura K, Kajiyama H, Kano H, Kikkawa F and Hori M 2011 Plasma-activated medium selectively kills glioblastoma brain tumor cells by down-regulating a survival signaling molecule, AKT Kinase *Plasma Med.* **1** 265–77

[20] Guerrero-Preston R, Ogawa T, Uemura M, Shumulinsky G, Valle B L, Pirini F, Ravi R, Sidransky D, Keidar M and Trink B 2014 Cold atmospheric plasma treatment selectively targets head and neck squamous cell carcinoma cells *Int. J. Mol. Med.* **34** 941–6

[21] Tan X, Zhao S, Lei Q, Lu X, He G and Ostrikov K 2014 Single-cell-precision microplasma-induced cancer cell apoptosis *PLoS One* **9** e101299

[22] Kaushik N, Kumar N, Kim C H, Kaushik N K and Choi E H 2014 Dielectric barrier discharge plasma efficiently delivers an apoptotic response in human monocytic lymphoma *Plasma Processes Polym.* **11** 1175–87

[23] Panngom K, Baik K Y, Nam M K, Han J H, Rhim H and Choi E H 2013 Preferential killing of human lung cancer cell lines with mitochondrial dysfunction by nonthermal dielectric barrier discharge plasma *Cell Death Dis.* **4** e642
[24] Fridman 2014 Selective cytotoxicity of indirect nonequilibrium atmospheric pressure plasma against ovarian clear-cell carcinoma *Springerplus* **3** 398
[25] Volotskova O *et al* 2012 Targeting the cancer cell cycle by cold atmospheric plasma *Sci. Rep.* **2** 636
[26] Kim J Y, Ballato J, Foy P, Hawkins T, Wei Y, Li J and Kim S O 2011 Apoptosis of lung carcinoma cells induced by a flexible optical fiber-based cold microplasma *Biosens. Bioelectron.* **28** 333–8
[27] Trachootham D, Alexandre J and Huang P 2009 Targeting cancer cells by ROS-mediated mechanisms: a radical therapeutic approach? *Nat. Rev. Drug Discov.* **8** 579–91
[28] Wang J and Yi J 2008 Cancer cell killing via ROS: to increase or decrease, that is a question *Cancer Biol. Ther.* **7** 1875–84
[29] Cairns R A, Harris I, Mccracken S and Mak T W 2011 Cancer cell metabolism *Cold Spring Harb. Symp. Quant. Biol.* **76** 299–311
[30] Graves D B 2012 The emerging role of reactive oxygen and nitrogen species in redox biology and some implications for plasma applications to medicine and biology *J. Phys. D: Appl. Phys.* **45** 263001–42
[31] Yan D, Malyavko A, Wang Q, Lin L, Sherman J H and Keidar M 2021 Cold atmospheric plasma cancer treatment, a critical review *Appl. Sci.* **11** 7757
[32] Ja Kim S, Min Joh H and Chung T H 2013 Production of intracellular reactive oxygen species and change of cell viability induced by atmospheric pressure plasma in normal and cancer cells *Appl. Phys. Lett.* **103** 153705
[33] Kaushik N K, Kaushik N, Park D and Choi E H 2014 Altered antioxidant system stimulates dielectric barrier discharge plasma-induced cell death for solid tumor cell treatment *PLoS One* **9** e103349
[34] Ishaq M, Kumar S, Varinli H, Han Z J, Rider A E, Evans M D M, Murphy A B and Ostrikov K 2014 Atmospheric gas plasma-induced ROS production activates TNF-ASK1 pathway for the induction of melanoma cancer cell apoptosis *Mol. Biol. Cell* **25** 1523–31
[35] Kim S J and Chung T H 2016 Cold atmospheric plasma jet-generated RONS and their selective effects on normal and carcinoma cells *Sci. Rep.* **6** 20332
[36] López-Lázaro M 2007 Dual role of hydrogen peroxide in cancer: possible relevance to cancer chemoprevention and therapy *Cancer Lett.* **252** 1–8
[37] Lim S D, Sun C, Lambeth J D, Marshall F, Amin M, Chung L, Petros J A and Arnold R S 2005 Increased Nox1 and hydrogen peroxide in prostate cancer *Prostate* **62** 200–7
[38] Chen Q, Espey M G, Krishna M C, Mitchell J B, Corpe C P, Buettner G R, Shacter E and Levine M 2005 Pharmacologic ascorbic acid concentrations selectively kill cancer cells: action as a pro-drug to deliver hydrogen peroxide to tissues *Proc. Natl Acad. Sci. USA* **102** 13604–9
[39] Zheng J, Lou J R, Zhang X X, Benbrook D M, Hanigan M H, Lind S E and Ding W Q 2010 N-Acetylcysteine interacts with copper to generate hydrogen peroxide and selectively induce cancer cell death *Cancer Lett.* **298** 186–94
[40] Kumar N, Park J H, Jeon S N, Park B S, Choi E H and Attri P 2016 The action of microsecond-pulsed plasma-activated media on the inactivation of human lung cancer cells *J. Phys. D: Appl. Phys.* **49** 115401

[41] Kurake N, Tanaka H, Ishikawa K, Kondo T, Sekine M, Nakamura K, Kajiyama H, Kikkawa F, Mizuno M and Hori M 2016 Cell survival of glioblastoma grown in medium containing hydrogen peroxide and/or nitrite, or in plasma-activated medium *Arch. Biochem. Biophys.* **605** 102–8

[42] Boehm D, Heslin C, Cullen P J and Bourke P 2016 Cytotoxic and mutagenic potential of solutions exposed to cold atmospheric plasma *Sci. Rep.* **6** 21464

[43] Nguyen N H, Park H J, Yang S S, Choi K S and Lee J-S 2016 Anti-cancer efficacy of nonthermal plasma dissolved in a liquid, liquid plasma in heterogeneous cancer cells *Sci. Rep.* **6** 29020

[44] Saito K, Asai T, Fujiwara K, Sahara J, Koguchi H, Fukuda N, Suzuki-Karasaki M, Soma M and Suzuki-Karasaki Y 2016 Tumor-selective mitochondrial network collapse induced by atmospheric gas plasma-activated medium *Oncotarget* **7** 19910

[45] Duan J, Lu X and He G 2017 The selective effect of plasma activated medium in an *in vitro* co-culture of liver cancer and normal cells *J. Appl. Phys.* **121** 013302

[46] Sklias K, Sousa J S and Girard P 2020 Role of short- and long-lived reactive species on the selectivity and anti-cancer action of plasma treatment *in vitro Cancers* **13** 615

[47] Agre P, King L S, Yasui M, Guggino W B, Ottersen O P, Fujiyoshi Y, Engel A and Nielsen S 2002 Aquaporin water channels–from atomic structure to clinical medicine *J. Physiol.* **542** 3–16

[48] Wu B and Beitz E 2007 Aquaporins with selectivity for unconventional permeants *Cell. Mol. Life Sci.* **64** 2413–21

[49] Bienert G P and Chaumont F 2014 Aquaporin-facilitated transmembrane diffusion of hydrogen peroxide *Biochim. Biophys. Acta, Gen. Subj.* **1840** 1596–604

[50] Bienert G P, Schjoerring J K and Jahn T P 2006 Membrane transport of hydrogen peroxide *Biochim. Biophys. Acta Biomembr.* **1758** 994–1003

[51] Miller E W, Dickinson B C and Chang C J 2010 Aquaporin-3 mediates hydrogen peroxide uptake to regulate downstream intracellular signaling *Proc. Natl Acad. Sci. USA* **107** 15681–6

[52] Almasalmeh A, Krenc D, Wu B and Beitz E 2014 Structural determinants of the hydrogen peroxide permeability of aquaporins *FEBS J.* **281** 647–56

[53] Rodrigues C, Pimpão C, Mósca A F, Coxixo A S, Lopes D, da Silva I V, Pedersen P A, Antunes F and Soveral G 2019 Human aquaporin-5 facilitates hydrogen peroxide permeation affecting adaption to oxidative stress and cancer cell migration *Cancers (Basel)* **11** 932

[54] Dalla S D F V, Zambonin L, Fiorentini D, Rizzo B, Caliceti C, Landi L, Hrelia S and Beitz E 2014 Specific aquaporins facilitate Nox-produced hydrogen peroxide transport through plasma membrane in leukaemia cells *Biochim. Biophys. Acta—Mol. Cell Res.* **1843** 806–14

[55] Ishibashi K and Sei S 1997 Aquaporin water channels in mammals *Clin. Exp. Nephrol.* **1** 247–53

[56] Papadopoulos M C and Saadoun S 2015 Key roles of aquaporins in tumor biology *Biochim. Biophys. Acta Biomembr.* **1848** 2576–83

[57] Abir-Awan M, Kitchen P, Salman M M, Conner M T, Conner A C and Bill R M 2019 Inhibitors of mammalian aquaporin water channels *Int. J. Mol. Sci.* **20** 1589

[58] Yan D, Xiao H, Zhu W, Nourmohammadi N, Zhang L G, Bian K and Keidar M 2017 The role of aquaporins in the anti-glioblastoma capacity of the cold plasma-stimulated medium *J. Phys. D: Appl. Phys.* **50** 055401

[59] MatÉs J É M, Pérez-Gómez C and De Castro I N 1999 Antioxidant enzymes and human diseases *Clin. Biochem.* **32** 595–603
[60] Cox A G, Winterbourn c.c and Hampton M B 2010 Mitochondrial peroxiredoxin involvement in antioxidant defence and redox signalling *Biochemical J.* **425** 313–25
[61] Glorieux C, Zamocky M, Sandoval J M, Verrax J and Calderon P B 2015 Regulation of catalase expression in healthy and cancerous cells *Free Radic. Biol. Med.* **87** 84–97
[62] Hasegawa Y, Takano T, Miyauchi A, Matsuzuka F, Yoshida H, Kuma K and Amino N 2003 Decreased expression of catalase mRNA in thyroid anaplastic carcinoma *Jpn. J. Clin. Oncol.* **33** 6–9
[63] Kaynar H, Meral M, Turhan H, Keles M, Celik G and Akcay F 2005 Glutathione peroxidase, glutathione-S-transferase, catalase, xanthine oxidase, Cu–Zn superoxide dismutase activities, total glutathione, nitric oxide, and malondialdehyde levels in erythrocytes of patients with small cell and non-small cell lung cancer *Cancer Lett.* **227** 133–9
[64] Bekeschus S *et al* 2020 xCT (SLC7A11) expression confers intrinsic resistance to physical plasma treatment in tumor cells *Redox Biol.* **30** 101423
[65] Yan D, Sherman J H, Canady J, Trink B and Keidar M The correlation between the cytotoxicity of cold atmospheric plasma and the extracellular H_2O_2-scavenging rate *IEEE Trans. Radiat. Plasma Med. Sci.* **2** 618–23
[66] Ma Y, Ha C S, Hwang S W, Lee H J, Kim G C, Lee K W and Song K 2014 Non-thermal atmospheric pressure plasma preferentially induces apoptosis in p53-mutated cancer cells by activating ROS stress-response pathways *PLoS One* **9** e91947
[67] Hanahan D and Weinberg R A 2011 Review hallmarks of cancer: the next generation *Cell* **144** 646–74
[68] Liu B, Chen Y and Clair D K S 2008 ROS and p53: a versatile partnership *Free Radic. Biol. Med.* **44** 1529–35
[69] Sablina A A, Budanov A V, Ilyinskaya G V, Agapova L S, Kravchenko J E and Chumakov P M 2005 The antioxidant function of the p53 tumor suppressor *Nature Med.* **11** 1306–13
[70] Vousden K H and Ryan K M 2009 p53 and metabolism *Nat. Rev. Cancer* **9** 691–700
[71] Borrás C, Gómez-cabrera M C and Viña J 2011 The dual role of p53: DNA protection and antioxidant *Free Radic. Res.* **45** 643–52
[72] Kang M Y, Kim H, Piao C, Lee K H, Hyun J W, Chang I and You H J 2013 The critical role of catalase in prooxidant and antioxidant function of p53 *Cell Death Differ.* **20** 117–29
[73] Yusupov M, Van der Paal J, Neyts E C and Bogaerts A 2017 Synergistic effect of electric field and lipid oxidation on the permeability of cell membranes *Biochim. Biophys. Acta—Gen. Subj.* **1861** 839–47
[74] Van der Paal J, Neyts E C, Verlackt C C W and Bogaerts A 2016 Effect of lipid peroxidation on membrane permeability of cancer and normal cells subjected to oxidative stress *Chem. Sci.* **7** 489–98
[75] Van der Paal J, Verheyen C, Neyts E C and Bogaerts A 2017 Hampering effect of cholesterol on the permeation of reactive oxygen species through phospholipids bilayer: possible explanation for plasma cancer selectivity *Sci. Rep.* **7** 39526
[76] Shinitzky M 1984 Membrane fluidity in malignancy adversative and recuperative *Biochim. Biophys. Acta—Rev. Cancer* **738** 251–61
[77] Bogaerts A, Yusupov M, Razzokov J and Van der Paal J 2019 Plasma for cancer treatment: how can RONS penetrate through the cell membrane? Answers from computer modelling *Front. Chem. Sci. Eng.* **13** 253–63

[78] Bauer G, Sersenová D, Graves D B and Machala Z 2019 Dynamics of singlet oxygen-triggered, RONS-based apoptosis induction after treatment of tumor cells with cold atmospheric plasma or plasma-activated medium *Sci. Rep.* **9** 13931

[79] Bauer G The synergistic effect between hydrogen peroxide and nitrite, two long-lived molecular species from cold atmospheric plasma, triggers tumors to induce their own cell death *Redox Biol.* **26** 101291

[80] Bauer G and Graves D B 2016 Mechanisms of selective antitumor action of cold atmospheric plasma-derived reactive oxygen and nitrogen species *Plasma Processes Polym.* **13** 1157–78

[81] Bauer G 2016 The antitumor effect of singlet oxygen *Anticancer Res.* **36** 5649–64

[82] Ahn H J, Kim K I, Hoan N N, Kim C H, Moon E, Choi K S, Yang S S and Lee J S 2014 Targeting cancer cells with reactive oxygen and nitrogen species generated by atmospheric-pressure air plasma *PLoS One* **9** e86173

[83] Yan D, Talbot A, Nourmohammadi N, Cheng X, Canady J, Sherman J and Keidar M 2015 Principles of using cold atmospheric plasma stimulated media for cancer treatment *Sci. Rep.* **5** 18339

[84] Yokoyama M, Johkura K and Sato T 2014 Gene expression responses of HeLa cells to chemical species generated by an atmospheric plasma flow *Biochem. Biophys. Res. Commun.* **450** 1266–71

[85] Yan D, Nourmohammadi N, Bian K, Murad F, Sherman J H and Keidar M 2016 Stabilizing the cold plasma-stimulated medium by regulating medium's composition *Sci. Rep.* **6** 26016

[86] Adachi T, Nonomura S, Horiba M, Hirayama T, Kamiya T, Nagasawa H and Hara H 2016 Iron stimulates plasma-activated medium-induced A549 cell injury *Sci. Rep.* **6** 20928

[87] Adachi T, Tanaka H, Nonomura S, Hara H, Kondo S I and Hori M 2015 Plasma-activated medium induces A549 cell injury via a spiral apoptotic cascade involving the mitochondrial-nuclear network *Free Radic. Biol. Med.* **79** 28–44

[88] Bekeschus S, Kolata J, Winterbourn C, Kramer A, Turner R, Weltmann K D, Br B and Masur K 2014 Hydrogen peroxide: a central player in physical plasma-induced oxidative stress in human blood cells *Free Radic. Res.* **48** 542–9

[89] Lindig B A, Rodgers M A J and Schaaplc A P 1980 Determination of the lifetime of singlet oxygen in D20 using 9,10-anthracenedipropionic acid, a water-soluble probe *J. Am. Chem. Soc.* **102** 5590–3

[90] Malyavko A, Yan D, Wang Q, Klein A L, Patel K C, Sherman J K and Keidar M 2020 Cold atmospheric plasma cancer treatment, direct versus indirect approaches *Mater. Adv.* **1** 1494–505

[91] Yan D, Talbot A, Nourmohammadi N, Sherman J H, Cheng X and Keidar M 2015 Toward understanding the selective anticancer capacity of cold atmospheric plasma—a model based on aquaporins *Biointerphases* **10** 040801

[92] Vandamme M *et al* 2012 ROS implication in a new antitumor strategy based on non-thermal plasma *Int. J. Cancer* **130** 2185–94

[93] Han X, Klas M, Liu Y, Sharon Stack M and Ptasinska S 2013 DNA damage in oral cancer cells induced by nitrogen atmospheric pressure plasma jets *Appl. Phys. Lett.* **102** 233703

[94] Kalghatgi S, Kelly C M, Cerchar E, Torabi B, Alekseev O, Fridman A, Friedman G and Azizkhan-Clifford J 2011 Effects of non-thermal plasma on mammalian cells *PLoS One* **6** e16270

[95] Kim G J, Kim W, Kim K T and Lee J K 2010 DNA damage and mitochondria dysfunction in cell apoptosis induced by nonthermal air plasma *Appl. Phys. Lett.* **96** 021502
[96] Sancar A, Lindsey-Boltz L A, Ünsal-Kaçmaz K and Linn S 2004 Molecular mechanisms of mammalian DNA repair and the DNA damage checkpoints *Annu. Rev. Biochem.* **73** 39–85
[97] Norbury C J and Zhivotovsky B 2004 DNA damage-induced apoptosis *Oncogene* **23** 2797–808
[98] Poulsen H E, Prieme H and Loft S 1998 Role of oxidative DNA damage in cancer initiation and promotion *Eur. J. Cancer Prev.* **7** 9–16
[99] Hanahan D and Weinberg R A 2011 Hallmarks of cancer: the next generation *Cell* **144** 646–74
[100] Bartkova J *et al* 2005 DNA damage response as a candidate anti-cancer barrier in early human tumorigenesis *Nature* **434** 864–70
[101] Halazonetis T D, Gorgoulis V G and Bartek J 2008 An oncogene-induced DNA damage model for cancer development *Science* **319** 1352–5
[102] Hoeijmakers J H J 2009 DNA Damage, aging, and cancer *N. Engl. J. Med.* **361** 1475–85
[103] Hać A, Brokowska J, Rintz E, Bartkowski M, Węgrzyn G and Herman-Antosiewicz A 2020 Mechanism of selective anticancer activity of isothiocyanates relies on differences in DNA damage repair between cancer and healthy cells *J. Nutr.* **59** 1421–32
[104] Lin L, Yan D, Gjika E, Sherman J H and Keidar M 2019 Atmospheric plasma meets cell: plasma tailoring by living cells *ACS Appl. Mater. Interfaces* **11** 30621–30
[105] Szatrowski T P and Nathan C F 1991 Production of large amounts of hydrogen peroxide by human tumor cells *Cancer Res.* **51** 794–8
[106] Yan D, Cui H, Zhu W, Talbot A, Zhang L G, Sherman J H and Keidar M 2017 The strong cell-based hydrogen peroxide generation triggered by cold atmospheric plasma *Sci. Rep.* **7** 10831
[107] Yan D, Lin L, Xu W, Nourmohammadi N and Jonathan H 2018 universality of micromolar-level cell-based hydrogen peroxide generation during direct cold atmospheric plasma treatment *Plasma Med.* **8** 335–43
[108] Nasri Z, Bruno G, Bekeschus S, Weltmann K D, von Woedtke T and Wende K 2021 Development of an electrochemical sensor for *in situ* monitoring of reactive species produced by cold physical plasma *Sens. Actuators B. Chem.* **326** 129007
[109] Keidar M, Yan D and Sherman J H 2019 *Cold Plasma Cancer Therapy* (Morgan & Claypool Publishers)
[110] Yan D, Wang Q, Adhikari M, Malyavko A, Lin L, Zolotukhin D B, Yao X, Kirschner M, Sherman J H and Keidar M 2020 A physically triggered cell death via transbarrier cold atmospheric plasma cancer treatment *ACS Appl. Mater. Interfaces* **12** 34548–63
[111] Wang Q, Malyavko A, Yan D, Lamanna O K, Hsieh M H, Sherman J and Keidar M 2020 A comparative study of cold atmospheric plasma treatment, chemical versus physical strategy *J. Phys. D: Appl. Phys.* **54** 095207
[112] Yan D, Wang Q, Malyavko A, Zolotukhin D B, Adhikari M, Sherman J H and Keidar M 2020 The anti-glioblastoma effect of cold atmospheric plasma treatment: physical pathway v.s. chemical pathway *Sci. Rep.* **10** 11788
[113] Tanaka S, Louis D N, Curry W T, Batchelor T T and Dietrich J 2013 Diagnostic and therapeutic avenues for glioblastoma: no longer a dead end? *Nat. Rev. Clin. Oncol.* **10** 14–26
[114] Holland E C 2000 Glioblastoma multiforme: the terminator *Proc. Natl Acad. Sci. USA* **97** 6242–4

[115] Weller M, Fisher B, Taphoorn M J B, Belanger K, Brandes A A, Marosi C, Bogdahn U, Curschmann J and Janzer R C 2005 Radiotherapy plus concomitant and adjuvant temozolomide for glioblastoma *N. Engl. J. Med.* **352** 987–96

[116] Chinot O L, Wick W, Mason W, Henriksson R, Saran F, Nishikawa R, Hilton M, Sc M, Abrey L and Cloughesy T 2014 Bevacizumab plus radiotherapy—temozolomide for newly diagnosed glioblastoma *N. Engl. J. Med.* **370** 709–22

[117] Yao X, Lin L, Soni V, Gjika E, Sherman J H, Yan D and Keidar M 2020 Sensitization of glioblastoma cells to temozolomide by a helium gas discharge tube *Phys. Plasmas* **27** 114502

[118] Yan D, Xu W, Yao X, Lin L, Sherman J H and Keidar M 2018 The cell activation phenomena in the cold atmospheric plasma cancer treatment *Sci. Rep.* **8** 15418

[119] Yan D, Lin L, Xu W and Nourmohammadi N 2019 Cold plasma-based control of the activation of pancreatic adenocarcinoma cells *J. Phys. D: Appl. Phys.* **52** 445202

Michael Keidar, Dayun Yan and Li Lin

Chapter 8

Animal studies and immune responses

Animal studies are critical for plasma medicine. Over the past decade, many *in vivo* studies have demonstrated the significant anti-cancer effect of cold atmospheric plasma (CAP) treatment over a series of xenografted tumor models in animals, and these works provide hope for using CAP as a clinical anti-cancer modality. The promising finding in animal studies also demonstrates the general safety of using CAP sources. Chapter 8 focuses on the research status and main conclusions obtained so far. Besides, as an immerging hot direction, immune responses due to CAP treatment have been introduced, which is not only a candidate to explain observed anti-cancer mechanism but also the cornerstone of CAP-based immunotherapy.

8.1 Cutaneous CAP treatment

Currently, the anti-cancer mechanism of CAP treatment *in vitro* is largely based on the increase of ROS via intracellular pathways. In addition, the aqueous environment, which facilitates the transformation of reactive species from the gas phase into the liquid phase, is also a key factor for the efficacy of CAP treatment. However, the current conclusions based on *in vitro* studies cannot be directly used to explain or predict the performance in animal studies. For example, the interface between the bulk plasma and tumor tissue is a relatively dry skin barrier, essentially different from the basic set of nearly all *in vitro* studies.

Animal studies tend to answer the most critical question in plasma medicine: how can CAP be used in the clinical setting to cure patients with tumors? Animal studies have gradually become the primary approach to demonstrate anti-cancer efficacy and discover novel tissue responses in current plasma medicine. To date, three types of animal models have been widely used in plasma medicine: subcutaneous,

doi:10.1088/978-0-7503-5537-7ch8

intraperitoneal, and orthotopic models [1]. Subcutaneous models provided the earliest and the most apparent feasibility demonstration of using CAP as an anti-tumor modality. In plasma medicine, most animal studies used subcutaneous models.

Compared with the abundant studies *in vitro*, animal studies are limited to dozens of preliminary experiments. So far, several groups have tested the anti-cancer effect of direct CAP treatment on subcutaneous xenograft tumors and melanoma in mice (table 8.1). CAP treatments were mainly performed by treating the skin above the tumor sites. Few studies applied the micro-size μ-CAP device to guide the tiny CAP jet, affecting the exposed tumorous tissues underneath the skin or the skull [2, 3]. Tumor volume measurement, bioluminescence imaging (BLI), and survival rate curve have been widely used to assess the anti-cancer effect of CAP treatment. All these studies achieved a similar but promising result; that is, a direct CAP treatment could significantly halt tumor growth in mice.

The earliest experiments were performed by Marc Vandamme *et al.* They used the glioblastoma U87MG tumor mouse xenograft model to test the anti-cancer effect of CAP treatment [5]. Their pioneering work demonstrated a floating electrode DBD treatment above the skin of a subcutaneous tumor could achieve

Table 8.1. Representative *in vivo* demonstrations on subcutaneous xenografted tumor models (2010–2018). Reproduced from [4] CC BY 4.0.

Reference	Years	Tumor types	Tumor size	Survival rate	Tumor diagnostics
[5]	2010	Glioblastoma	Decreased	N/A	Bioluminescence imaging
[6]	2010	Bladder cancer	Decreased	Increased	Tissue size measurement
[7]	2011	Glioblastoma	Decreased	Increased	Bioluminescence imaging
[8]	2012	Pancreatic carcinoma	Decreased	N/A	Bioluminescence imaging
[9]	2012	Glioblastoma	Decreased	N/A	Bioluminescence imaging
[10]	2013	Neuroblastoma	Decreased	Increased	Tissue size measurement
[11]	2014	Melanoma	Decreased	N/A	Tissue size measurement
[12]	2014	Head and neck cancer	Decreased	N/A	Tissue size measurement
[13]	2015	Melanoma	Decreased	Increased	Tissue size measurement
[14]	2015	Endometrioid adenocarcinoma	Decreased	N/A	Tissue size measurement
[15]	2016	Glioblastoma	Decreased	N/A	Tissue size measurement
[2]	2016	Breast cancer	Decreased	N/A	Tissue size measurement
[16]	2017	Melanoma	Decreased	N/A	Bioluminescence imaging
[17]	2018	Colorectal tumor	Decreased	N/A	Tissue size measurement

a promising half decrease in tumor volume in the treated mice. Corresponding mouse length of survival increased about 60% after the FE-DBD treatment. In this study, the authors also discovered a significant trend that the five-consecutive plasma treatments would cause a better anti-cancer effect than a long single plasma treatment. A multi-time consecutive treatment may consecutively trigger some long-term anti-tumor effects, such as immune responses. In the same year, Michael Keidar *et al* performed an animal study on a bladder tumor xenograft model on mice through the direct CAP jet treatment (figure 8.1(a)) [6]. After just 2 min of treatment, the tumor significantly decreased its size and could not be observed via gross inspection after 24 h. These two preliminary studies demonstrate the promising potential of the CAP as an anti-cancer method with no toxic side effects and physical damage.

Due to the subcutaneous nature of melanomas, it has become one of the most promising candidates for CAP cancer therapy. Fortunately, the preliminary tests on the melanoma model achieved very promising results [18]. The earliest test was done by Michael Keidar *et al* in 2011 [6]. They found that the CAP treatment nearly wholly inhibited tumor growth in a murine xenografted melanoma model over 3 weeks post treatment (figure 8.1(b)) [6]. Correspondingly, the mice survival length has been drastically extended. Through the pathological analysis, Natalie Chernets *et al* confirmed that the CAP-treated melanoma tissue was completely absent on the 22nd day after the treatment [13]. Correspondingly, the mice survival rate increased from 0% to 66.7%, roughly a month succeeding the treatment [13].

Using bioluminescence imaging approach, another later investigation observed the medium level of tumor size decrease after CAP treatment [16]. They compared the anti-cancer effect of the single weekly treatment and the fractionated weekly

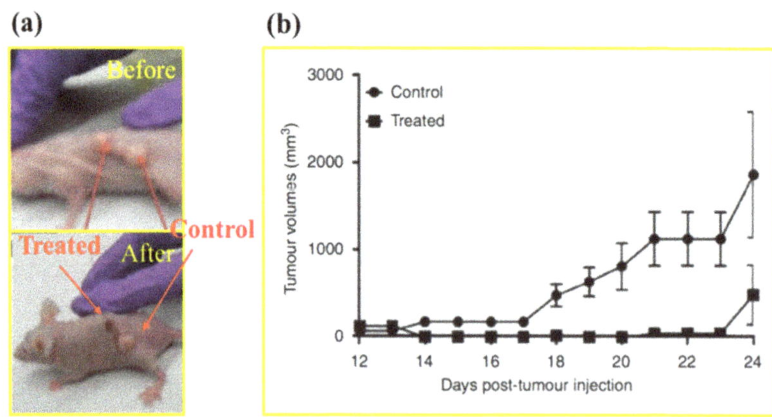

Figure 8.1. An early demonstration of cutaneous treatment using subcutaneous models. (a) Mice with subcutaneous xenografted bladder tumors before and 24 h post-atmospheric pressure plasma jet (APPJ) treatment [6]. (b) Tumor volumes in a murine melanoma model. Reprinted from [6], copyright (2011), with permission from Springer Nature.

treatment. They found that the group using one treatment per week caused a better anti-cancer effect on melanoma than the group using five fractionated treatments per week when the whole treatment was kept the same [16]. This trend contradicts the finding in the first test *in vivo* [7]. Currently, the conclusion that a fractionated but shorter treatment is better than a single but longer CAP treatment is still questionable. However, the fractionated, prolonged treatment definitely should be recommended in future animal studies and clinical trials.

8.2 Intracranial models

So far, the CAP-based anti-cancer capability was mainly demonstrated via subcutaneous models. However, in many other cases, an in-depth tumor will not be directly exposed to bulk plasma except when surgery is performed before CAP treatment. CAP-activated solutions (PAS) show more practical feasibility than direct treatment for intracranial models of deep tumors. A series of animal studies has demonstrated the injection of PAS to affect tumor in-depth (table 8.2).

Fumi Utsumi *et al* first demonstrated that injecting plasma-activated Ringer's lactate into the subcutaneous tumors from xenografted chemical-resistant ovarian cancer cells drastically inhibited tumor growth in mice [19]. Ringer's lactate solution is an isotonic solution containing sodium chloride, sodium lactate, potassium chloride, calcium chloride, as well as sodium bicarbonate to buffer. Currently, Ringer's lactate solution is closer than PBS to being translated and used as a cancer modality [24–26]. Hiromasa Tanaka *et al* demonstrated the noticeable anti-cancer effects of CAP-activated Ringer's lactate solution (PAL) after injecting it into subcutaneously xenografted human cervical cancer SiHa tumors in mice [20]. This method effectively reduced tumor volumes and had no apparent adverse effects on the mice (figure 8.2) [20]. PAS can also be used to treat intraperitoneally xenografted tumor models. Shigeomi Takeda *et al* demonstrated that PAM effectively decreased

Table 8.2. Examples of using PAS in cancer treatment *in vivo*. DBD: dielectric barrier discharge; PBS: phosphate buffered saline. We enumerated studies in mice using the biologically adaptable solutions here.

Reference	Years	Plasmas	Solutions	Cancer cells	Tumor models	Treatment
[19]	2013	Plasma jet	Medium	Ovarian cancer (NOS2/2TR)	Subcutaneously xenografted	Injection
[20]	2016	Plasma jet	Ringer's solution	Cervical cancer (SiHa)	Subcutaneously xenografted	Injection
[21]	2017	Plasma jet	Medium	Ovarian cancer (ES_2)	Intraperitoneally xenografted	Injection
[22]	2017	Plasma jet	Medium	Gastric cancer (GCIY-EGFP)	Intraperitoneally xenografted	Injection
[23]	2017	Plasma jet	Medium	Pancreatic cancer (6606PDA)	Intraperitoneally xenografted	Injection

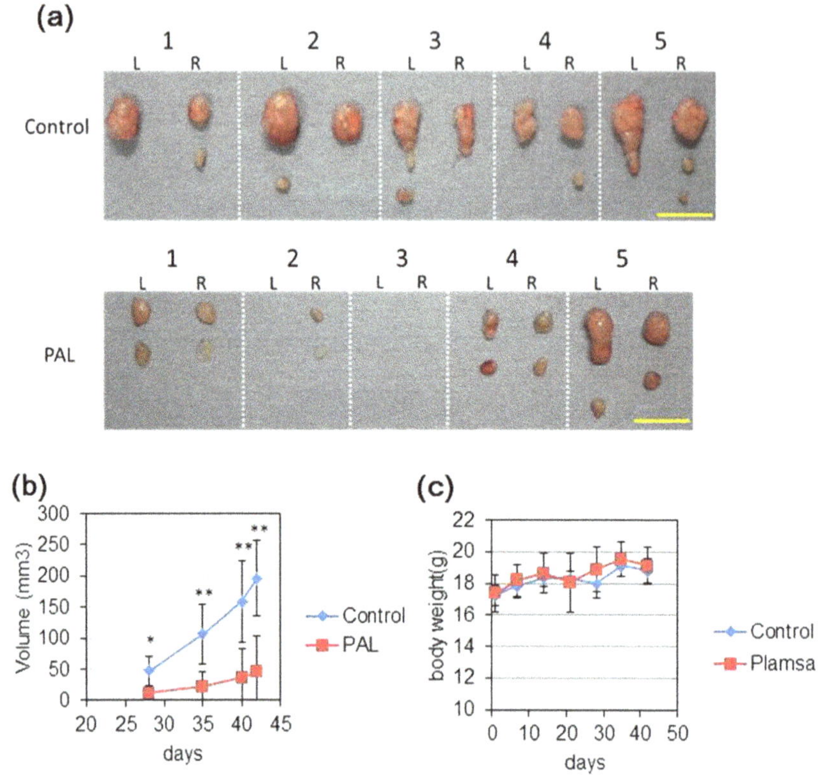

Figure 8.2. Anti-tumor effects of PAL in mice injected with a human cervical cancer cell line SiHa [20]. Female Balb/c nude mice inoculated with 1.5×10^3 SiHa cells with Matrigel in both hind flanks were administered 200L of PAL, or Lactec as a control, 3 times weekly for 6 weeks. (a) Representative tumor images at the study endpoint, day 42. location of tumors: left- (L) or right- (R) side hind flank, respectively. Yellow bars: 1 cm. (b) Starting on day 28 after inoculation, tumor size was calculated once a week. (c) Body weight was measured once a week. Reproduced from [20]. CC BY 4.0.

the formation of peritoneal metastatic nodules by 60% without causing observable adverse events [22].

Similar anti-cancer performance by injection of PAS has been reported in subsequent studies (figure 8.3) [23, 27]. Tumor nodes in the peritoneum were imaged using magnetic resonance imaging (MRI), and obviously, nodes drastically decreased in the PAM-treated group (II). Here, the PAM used was an NTP-treated medium. MRI-based tumor volume calculation showed a 50% reduction in total tumor growth after PAM treatment. The entire tumor mass drastically decreased ~40% in the PAM-treated group when animals were sacrificed on day 28 (d28) [23].

Current translations to the clinic are promising. PAS can be used in the lavage in patients suffering from peritoneal carcinomatosis, adjuvant to standard chemotherapy [28, 29]. PAS can also synergistically enhance the therapeutic efficacy of chemotherapy drugs and other chemicals [30–32]. More importantly, current studies

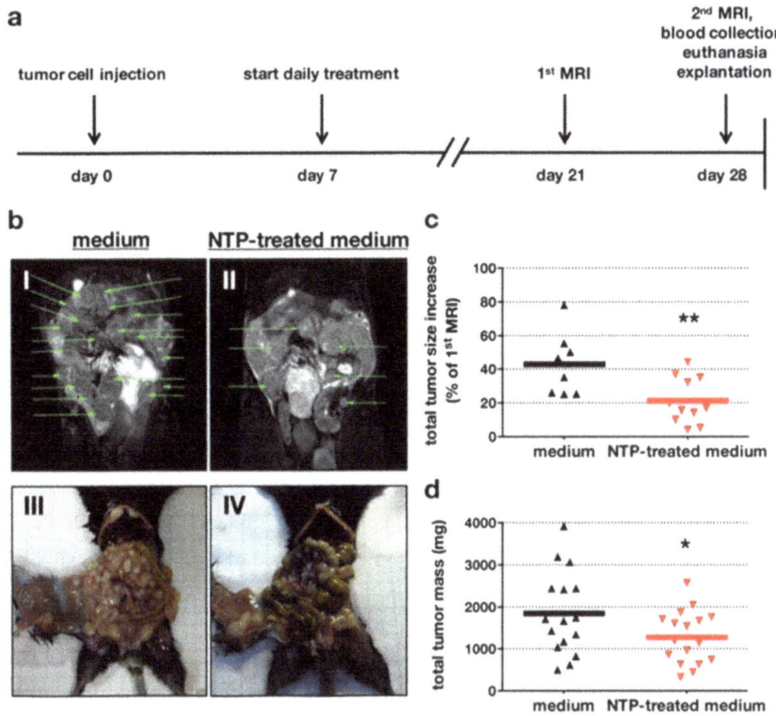

Figure 8.3. PAM significantly decreased the number, growth, and volume of pancreatic lesions *in vivo* [23]. (a) Experimental timeline. (b) Representative tumor imaging. I, II: 2nd MRI was performed at the end of treatment (d28). Tumor nodes were indicated by green arrows. III, IV: Macroscopic dissection imaging of control (III) and PAM-treated animals (IV). (c) Tumor volume based on MRI imaging. (d) Total tumor mass of the sacrificed animals on day 28. Reproduced from [23].

on immuno-deficient nude mice have shown that oral lavage treatment of CAP-activated water did not cause lethal effects or acute toxicity [33, 34]. These mice had no significant changes in survival status, body weight, organ coefficient, function, and tissue structure of the liver, spleen, heart, lung, and kidney [33, 34]. Despite the limited number of studies *in vivo*, PAS has shown promising results in clinical cancer therapy, especially by injection at specific tumor sites.

Recently, a novel approach to using PAS in cancer therapy has been proposed. Instead of direct inhibition of tumor tissue, this approach focuses on suppressing the growth of residual tumor tissues or cells after surgery, which is a primary cause of relapse and progression. In one experiment, PAS was used to make a biogel to be used in post-surgical treatments. This fillable plasma-activated biogel was made on a thermosensitive biogel, (poly-dl-lactide)-(poly-ethylene-glycol)-(poly-dl-lactide), PLEL, with the aid of PAS for local post-surgical removal of tumors [35]. Preliminary studies demonstrated that the plasma-activated PLEL biogel (PAPB) completely eliminated residual *in situ* tumor tissue recurrence after removal surgery without showing systemic toxicity (figure 8.4). More importantly, PAPB expressed a slow release of ROS's feature. Together, this work provides the groundwork for using PAS in local post-operative cancer treatments in the future [35].

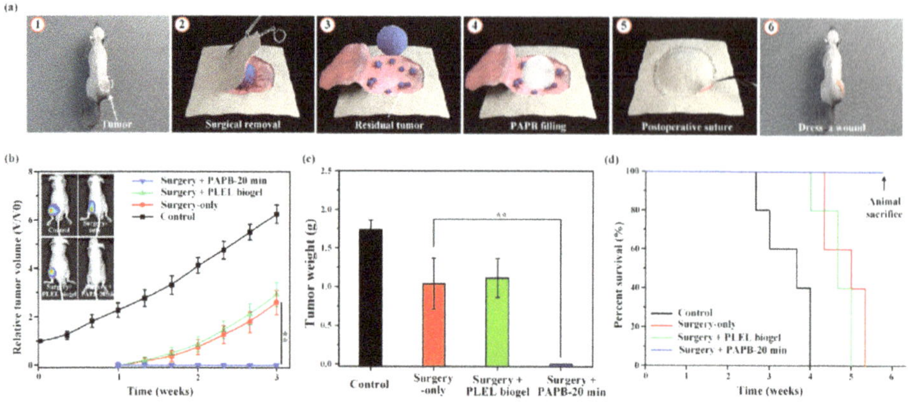

Figure 8.4. Anti-cancer performance of plasma-activated PLEL biogel (PAPB) on residual tumor after surgical removal [4, 35]. (a) Schematic illustration of the post-operative PAPB treatment. (b) Whole-body bioluminescence imaging and tumor growth in mice: control, surgical removal-only, surgical removal + PLEL (thermosensitive biogel), and surgical removal + PAPB treatment. (c) Excised tumors' weight. (d) Percent survival of tumor-bearing mice. Reprinted from [35], copyright (2021), with permission from Elsevier.

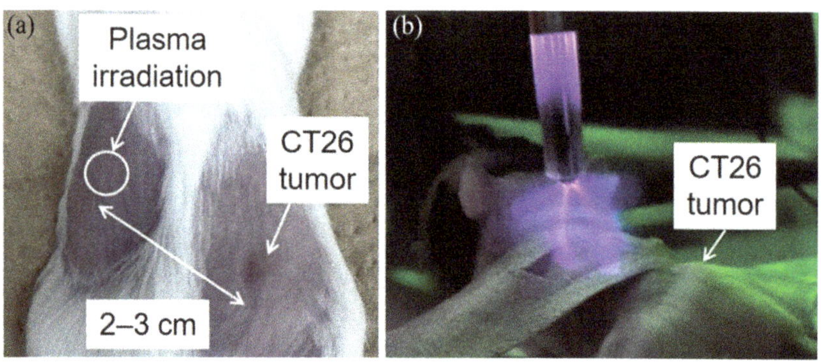

Figure 8.5. Abscopal anti-cancer effects of a CAP treatment on normal tissue near a tumor site [4]. (a) Treatment site on normal skin and the skin above murine colorectal carcinoma CT26 tumor on day 6 after treatment. (b) Treatment on the normal skin could affect the nearby tumor. Reproduced from [37]. CC BY 4.0.

8.3 Abscopal phenomena

Different from other anti-cancer modalities, CAP treatment also shows an abscopal effect in animal studies. Though only two studies have been published, they provide a novel route to use CAP in therapy. Unlike other works, these studies demonstrated that a CAP treatment on the nearby tumor site or the nearby healthy tissue suppressed tumor growth at a non-treated site on a mouse. And these unusual phenomena appeared even just one day after the treatment [36]. As shown in figure 8.5, when CAP treatment was performed on the skin just above the healthy tissue of the left limb, the abscopal effect would be significant only for the mice with tumors on the right limb [37]. And the abscopal effect on the skin above normal tissue was even comparable to the anti-tumor efficacy of a direct treatment on the skin above tumorous tissue. So far, the abscopal mechanism is unknown. We know that the innate immune response may trigger these rapid abscopal effects by producing inflammatory cytokine (IFN-γ) from splenocytes after CAP treatment [36].

The physical effect from CAP may be a good candidate to explain the abscopal anti-tumor phenomena. As we introduced in chapter 6, the electromagnetic (EM) effect in CAP may affect the target in an area much bigger than the plasma plume's size. For example, a typical APPJ with a tip size of less than 1 mm could affect an area of cells with a diameter of at least 2 cm based on the physical factors in CAP [38]. Regarding the abscopal phenomena that could affect the tumor by the treatment on another site about 2–3 cm away, the physical effect in CAP is a persuasive candidate. However, there is still a long way to go to confirm this speculation. Physical effect in CAP is much more complex than the well-understood chemical effects such as redox biological effect. The physical effect involves sophisticated spatial and energy distribution and probably needs to consider the anisotropic effect. All these factors may contribute to novel biological effects, such as abscopal phenomena.

8.4 Sensitization of tumor to drugs

In plasma medicine, the primary strategy is getting a direct anti-tumor effect through a treatment. Recently, we proposed a novel way to use CAP, which caught attention. We used CAP source to sensitize cancer cells to chemotherapeutic drugs. This strategy was inspired by our pioneering works on the activation phenomena *in vitro* [39–41], which has been introduced in chapter 4. Unlike previous studies focusing on subcutaneous tumors, our new approach focused on using physical factors such as the EM effect in CAP to affect tumors in-depth. Though the underlying mechanism is unknown, several demonstrations have been published.

Glioblastoma (GBM) is an aggressive and fatal brain cancer with a low survival rate, partially due to its strong resistance to current therapeutic approaches [42, 43]. Temozolomide (TMZ) is an FDA-approved alkylating chemotherapeutical agent. TMZ has been widely used in GBM treatment, particularly in high-grade malignant cases, though noticeable side effects commonly exist [43]. Our first trial demonstrated that a prototype helium gas discharge tube (PDT) can sensitize glioblastoma cells to the cytotoxicity of TMZ *in vitro* [44]. Another demonstration was using an improved radial helium cold plasma discharge tube as a tunable EM emission source (figure 8.6(a)). Compared to the normal astrocyte cell line hTERT/E6/E7, PDT selectively improved the cytotoxicity of TMZ on two glioblastoma cell lines, U87MG and A172 *in vitro* [45]. Preliminary animal studies demonstrated more than 100% improved mean

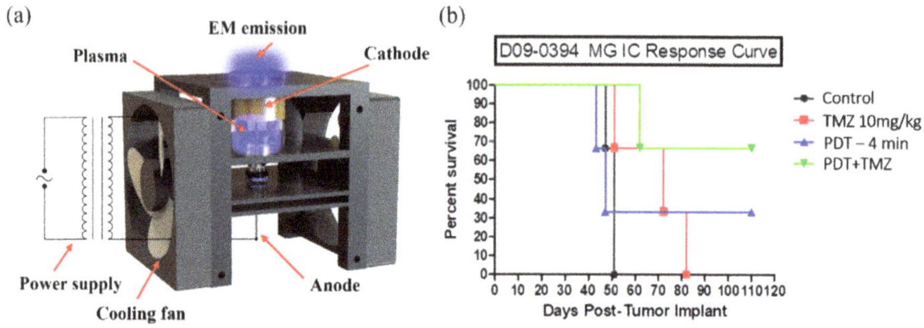

Figure 8.6. Sensitization of brain cancer cells to Temozolomide (TMZ) by a cold plasma discharge tube (PDT) [4]. (a) Structure of PDT source. (b) Mouse survival rate curve. Reproduced from [45]. CC BY 4.0.

survival day of patient-derived xenografted glioblastoma mice models (figure 8.6(b)). PDT is not only independent of extra helium supply but also capable of resisting environmental interference. Besides, PDT has the potential to be built in micro-scale [45]. In short, these two works were first to demonstrate the promising application of PDTs in GBM therapy as a non-invasive and portable modality. To date, the effective EM frequency range of sensitization effect is totally unknown, which is the main question to be answered in the subsequent studies.

In addition to plasma discharge tube, a CAP jet can also sensitize brain cancer cells to TMZ. CAP treatment was performed on the mouse head, in which U87MG-RedFluc cells were implanted intracranially and grew for one week [46]. CAP jet was used to treat the head of a mouse above the tumorous site for 1 min. The control group just used helium flow without discharge. This was immediately followed by the administration of TMZ or vehicle control for 5 days a week for 2 weeks. Bioluminescence imaging revealed that CAP treatment significantly sensitized brain tumors to the cytotoxicity of TMZ (figure 8.7). In contrast, helium flow alone did not show a similar effect. Specifically, a single non-invasive CAP pre-treatment on the mouse head with brain tumor resulted in ~70% bioluminescence decrease compared with the control group with helium flow, when all mice were given TMZ as the drug simultaneously. The combination of CAP with TMZ treatment effectively prevented GBM progression before the sacrifice of mice [46]. In short, these preliminary studies demonstrated that the physical effect of CAP could penetrate the skin and even the skull to sensitize brain tumors to TMZ, providing an unprecedented vision to use CAP as a non-invasive anti-cancer modality.

8.5 Transdermal anti-tumor mechanism

CAP-originated reactive species may dominate the anti-tumor mechanism of CAP treatment *in vivo* [4]. However, a single reactive species-based treatment does not generate the same anti-tumor effect as the CAP treatment does. For example, using the gel/H_2O_2 mixture did not generate a tumor killing efficacy as significant as that

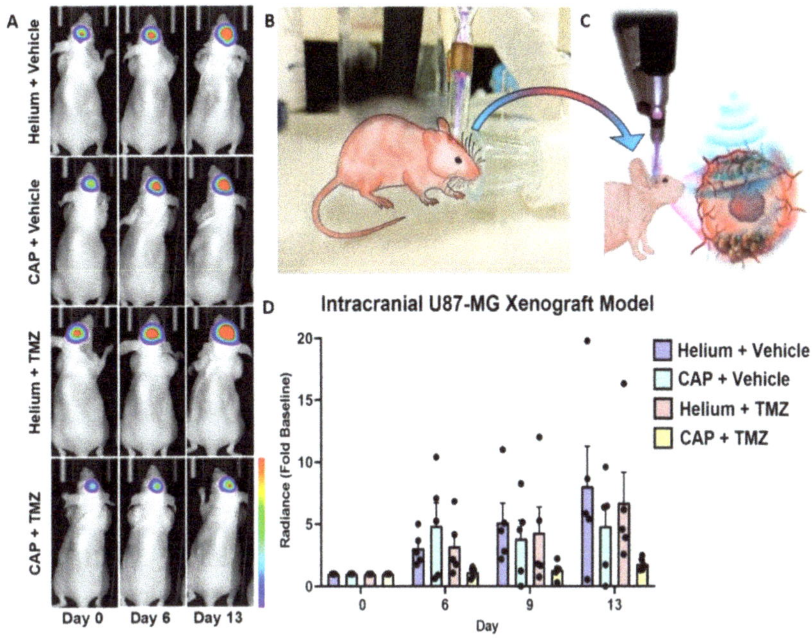

Figure 8.7. A CAP jet can sensitize brain tumor to TMZ by non-invasive treatment. (a) Representative bioluminescence imaging at baseline (day 0), day 6, and day 13 following a single non-invasive CAP treatment. (b and c) Schematic illustration of the non-invasive treatment on GBM *in vivo*. (d) Quantitative luminescence. Reprinted with permission from [46], copyright (2022) American Chemical Society.

of DBD did on the melanoma in the mouse [13]. The gel/H_2O_2 mixture led to delayed tumor growth, but the tumor eventually progressed through the treatment [13]. Based on this result, it is hypothesized that other reactive species may also impact tumorous tissue. The biological effect of H_2O_2 on tumorous tissues may be quite different from what is seen in *in vitro* studies, performed in a petri dish or multi-well plate setting.

In plasma medicine, ROS generated from CAP has been widely regarded as the main factor to cause cellular damage and ultimate cell death *in vitro* [47, 48]. The noticeable intracellular ROS rise is a representative cellular response to CAP treatment *in vitro* [18]. Such a common phenomenon may be due to the diffusion or the delivery of extracellular ROS into cells. However, there is no evidence to preclude the mechanism that complex biological pathways trigger the secondary ROS generation due to other factors. Interestingly, a similar ROS rise was also observed in the tumorous tissue after CAP but not helium flow treatment (figure 8.8) [49]. Thus, it is reasonable to speculate that the redox balance change is also a critical anti-tumor mechanism in animal studies.

What factors in CAP trigger these ROS rise phenomena? Due to the complex nature of CAP and subcutaneous tissues, such an anti-tumor mechanism is an open question. For subcutaneous models, CAP treatment was mainly performed by treating the skin just above tumorous tissues. Clearly, in such a setting, either

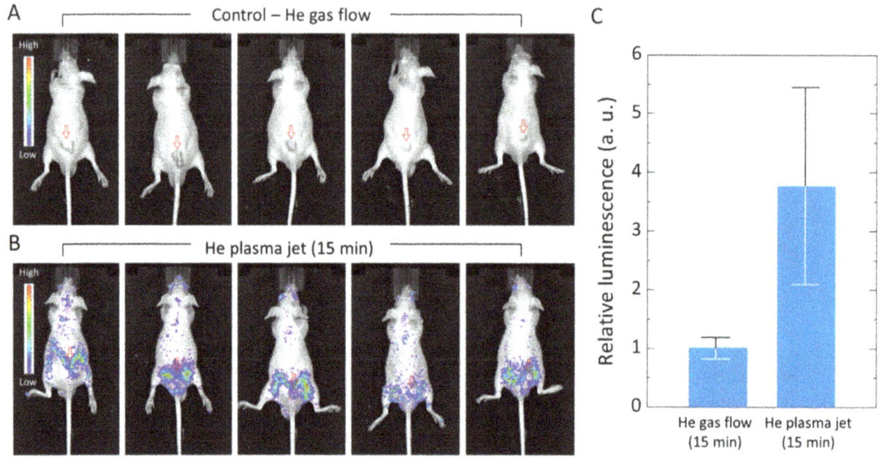

Figure 8.8. Detection of ROS after helium plasma jet treatment of a tumor on a live mouse. Spatial luminescence spatial distribution: (a) helium gas flow and (b) helium plasma jet treatment. (c) Total luminescence. Reproduced from [49]. Copyright IOP Publishing Ltd. All rights reserved.

chemical or physical factors in CAP must penetrate the skin barrier to trigger biological responses and pathways to further inhibit tumorous growth or completely eradicate the observable tumor tissues. These features give CAP treatment a potential non-invasive nature. The physical pathways in the skin for reactive species' transdermal diffusion or delivery are also unknown. Possible pathways in CAP treatment may include hair follicles, microneedles, other transcellular routes, and the electroporation effect [50].

So far, many attempts have been made to explain why an inhibition effect is present in a transdermal CAP treatment in subcutaneous tumor models. Generally, if we just consider chemical factors, reactive species need to penetrate skin or other tissues outside tumorous tissues [49, 50]. Here, complex interaction between reactive species and tumorous microenvironment may be involved. Experimental observations found a CAP treatment on pancreatic tumorous tissue only caused the observable apoptosis on the upper 3–5 cell layers [51]. Contact- and marker-independent Raman microscopy found that reactive species could penetrate the cervical epithelium sample's basal cell layer in about 270 μm [52]. Although reactive species may just penetrate subcutaneous tissue with a minimal depth, they may activate secondary messengers to finally cause a killing effect on tumorous tissues through complex pathways such as immune responses. So far, no obvious experimental evidence demonstrates the strong transdermal diffusion of reactive species such as ROS across the skin. A recent discovery of the physically-based killing effect provides another perspective to understand these puzzles. CAP's physical effect can penetrate a dielectric material with a thickness of more than 1 mm [53], close to human skin's thickness. Despite the material property of culture dishes or plates being distinct from human skin, a similar physical effect may exist when CAP is used in transdermal treatment.

Despite lacking direct evidence, preliminary studies have been performed by using some skin substitute materials, such as agarose gel, to simulate the potential transdermal diffusion of reactive species [50]. Though only a few studies were performed, some general trends have been found. The initial experiment set a 1.5 mm thick agarose gel between the plasma and deionized (DI) water [54]. H_2O_2 generation in DI water covered by agarose gel was time-dependent after the CAP treatment, suggesting it was gradually and slowly transported through the agarose gel. The whole transportation time lasted at least 30 min [54]. We know without the agarose gel barrier, reactive species will be generated in DI water instantly. In another experiment, the transportation of reactive species through the agarose film was found to be highly dependent on the film thickness [55]. In addition, a shorter distance between CAP source and the agarose film caused a stronger reactive species generation in the DI water underneath the film. Another important observation was that a 10 mm gas gap between the agarose film and the surface of deionized water could completely inhibit reactive species' transmembrane diffusion [55]. Similarly, reactive species' diffusion across a 1 mm thick protein-containing gelatin tissue-like film and protein-free carbon/oxygen poly(vinylalcohol) film into PBS was investigated [56]. It was found that incorporating bovine serum albumin in the tissue-like film weakened the ROS in the PBS underneath such a film [56]. Clearly, this study told us the reactive species' diffusion or delivery through the skin would also face many proteins and other reactive components, which becomes the natural barrier for the diffusion of reactive species. In short, these preliminary investigations suggest that the diffusion of reactive species across a mm-level tissue-like film is possible.

Recently, the transdermal diffusion of reactive species across animal skin, such as mouse and pig skin, was also studied. For example, an air plasma source was used to directly treat the mouse skin with a thickness of 0.75 mm. The authors did not observe the formation of H_2O_2 and NO_2^-/NO_3^- in the deionized water underneath the skin [57]. However, they indeed found the diffusion of NO_2^-/NO_3^- rather than H_2O_2 from CAP-treated DI water across the skin. Thus, the authors concluded that only these reactive species in aqueous solutions could penetrate the skin through some avenues, such as hair follicles or transcellular routes [57]. However, the transdermal diffusion of RON/RNS across pig skin was observed under specific discharge frequency conditions [58, 59]. A study found that a 300 ns 50 kV cm^{-1} pulsed electric field increased the reactive species transdermal diffusion across a pig skin model [60]. Altogether, these preliminary investigations suggest that CAP-originated reactive species' transdermal diffusion across a millimeter-level thick skin or skin mimic film is possible.

8.6 Thermal effect

To date, some studies found the temperature rise on the CAP-treated skin or tumor tissue is negligible and concluded that CAP is thermally safe for normal skin and tissue [16]. However, this conclusion may not be correct for some cases because the thermal effect of CAP treatment cannot be simply ignored for *in vivo* studies.

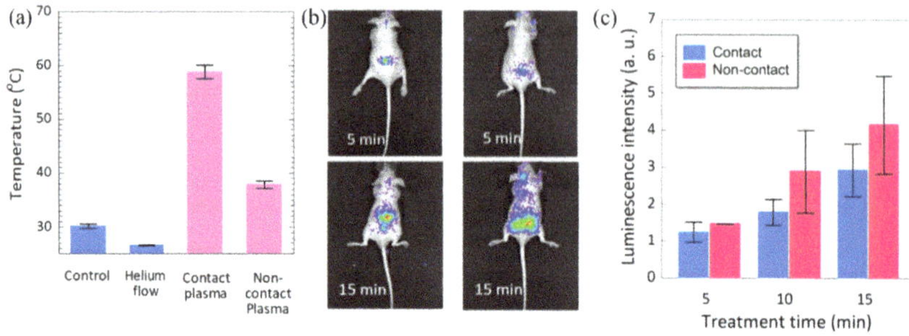

Figure 8.9. Non-contact CAP treatment caused a stronger biological response than contact treatment. Reprinted from [61]. CC BY 4.0. (a) Maximum temperature of mouse skin. Helium flow alone can cool the skin temperature because of the cooler temperature of helium gas in the pressurized tank. (b) *In vivo* ROS detection via luminescence from the reaction of luminol with ROS: left (contact treatment) and right (non-contact treatment). (c) The average luminescence intensity.

For example, in the earliest study, the subcutaneous temperature of a mouse was still close to body temperature after CAP treatment [5]. In another study, however, the highest temperature in the treated mouse skin was already higher than 50 °C after just 4 min of treatment [16]. Quite clearly, the heating effect in this case and other similar studies needs to be seriously considered. Unfortunately, the temperature and thermal effect data during a CAP treatment were absent in most studies. We recommend that this data be provided in future studies, particularly animal studies. Clearly, the name of 'CAP' cannot guarantee the corresponding plasma is 'cold' and 'nonthermal'. 'Hot' CAP sources are not cold plasma sources.

Thermal effects may have a complex biological impact on tumors. When CAP source was used to treat mice for 15 min, the maximum temperature on the skin could reach about 60 °C (figure 8.9(a)). Not surprisingly, when the plasma plume detached from the skin, the measured maximum temperature decreased by about 20 °C. Surprisingly, *in vivo* ROS detection revealed that contact CAP treatment caused a stronger ROS rise than non-contact CAP treatment did (figures 8.9(b) and (c)). In other words, the stronger thermal effect corresponded to a weaker ROS rise *in vivo* [61]. In short, the thermal effect from a CAP treatment is an important physical factor in animal studies, particularly when plasma is not really cold in many cases.

8.7 Immune responses

How do the CAP-generated reactive species affect subcutaneous tumor tissues? Immune responses provide clues to answer this critical question. Immune responses have been observed in plasma medicine for a while. One early representative observation found that macrophages could be activated *in vitro* by a uniform nanosecond pulsed DBD (nspDBD), which could improve the healing effect on an artificial wound [62]. In one of our recent studies, it was found that CAP could enhance an innate and adaptive immune system. A short exposure of CAP to macrophages and T-cells *in vitro* triggered an inflammatory phenotype leading to a better antigen-presenting and effector cell function, respectively [63].

A comprehensive analysis of the CAP triggered immune response was performed recently. Cutaneous melanoma is the most aggressive type of skin cancer. A representative study treated skin biopsies from human melanoma *ex vivo* with a CAP jet source (kINPen) [64]. In metastatic melanoma skin biopsies from six patients, the baseline release levels of 24 chemokines/cytokines were investigated (figure 8.10). Two general trends were observed. On one hand, CAP treatment could affect many chemokines or cytokines generation in different ways. It increased the generation of CCL2, CCL5, EGF, IL4, and IL17A. However, CAP treatment decreased the generation of IL6, IL10, PDGF, TNFα, and VEGF. On the other hand, these changing patterns were patent specified. These CAP triggered chemokines/cytokines change did not appear in roughly half of six patients [64].

CAP-based immune response can explain why transdermal CAP treatment can effectively inhibit subcutaneous tumorous growth. CAP may activate the immune system to attack tumorous tissue by reactive species or other factors in CAP [62, 63]. Vandana Miller *et al* have provided a comprehensive introduction about the promising cancer immunotherapy based on the CAP treatment [65]. The development of CAP in combination with cancer immunotherapeutics has received growing attention. Here, we just give a brief introduction. Abraham Lin *et al* named such immune response as the induction of immunogenic cancer cell death (ICD) [66]. The CAP treatment on cancer cells triggered cancer cells to emit signals known as damage-associated molecular patterns (DAMPs), which can attract and stimulate local immune cells (figure 8.11) [67]. One of such DAMP is calreticulin (CRT). Another study have shown that the CAP treatment on cancer cell lines caused the increasing exposure of CRT on the surface of cancer cells [68]. The specific molecular pattern signals such as adenosine triphosphate (ATP) may also be activated by CAP treatment, which further triggers the immunogenic attack on cancer cells [67]. Besides, the CAP treatment on the subcutaneous tumors also elicited the emission of danger signals and recruitment of antigen-presenting cells into tumors [17].

CAP causes ICD far beyond these approaches. For example, the injection of the CAP-treated CT26 colorectal cancer cells in mice caused a noticeable growth inhibition in the tumor compared with the case just injecting the CT26 cells without treatment [68]. In other words, the CAP-treated cancer cells were used as a whole-cell vaccine to elicit protective immunity in at least CT26 colorectal tumor models [69]. Similar approaches were reported in other studies [70]. Furthermore, the synergistic use of CAP with vaccination also enhanced the cancer-specific T-cell responses [68, 69]. Together, immunogenic cell death may be a critical mechanism underlying the anti-tumor capability of a CAP treatment.

Other mechanisms may trigger the immune response. We have identified a cell-based H_2O_2 generation during the direct CAP treatment *in vitro* [71]. This phenomenon may also exist *in vivo*. The cell-based H_2O_2 generation provides a novel mechanism to explain the potential immune response. The short-lived reactive species such as superoxide or single oxygen may activate cancerous cells to generate

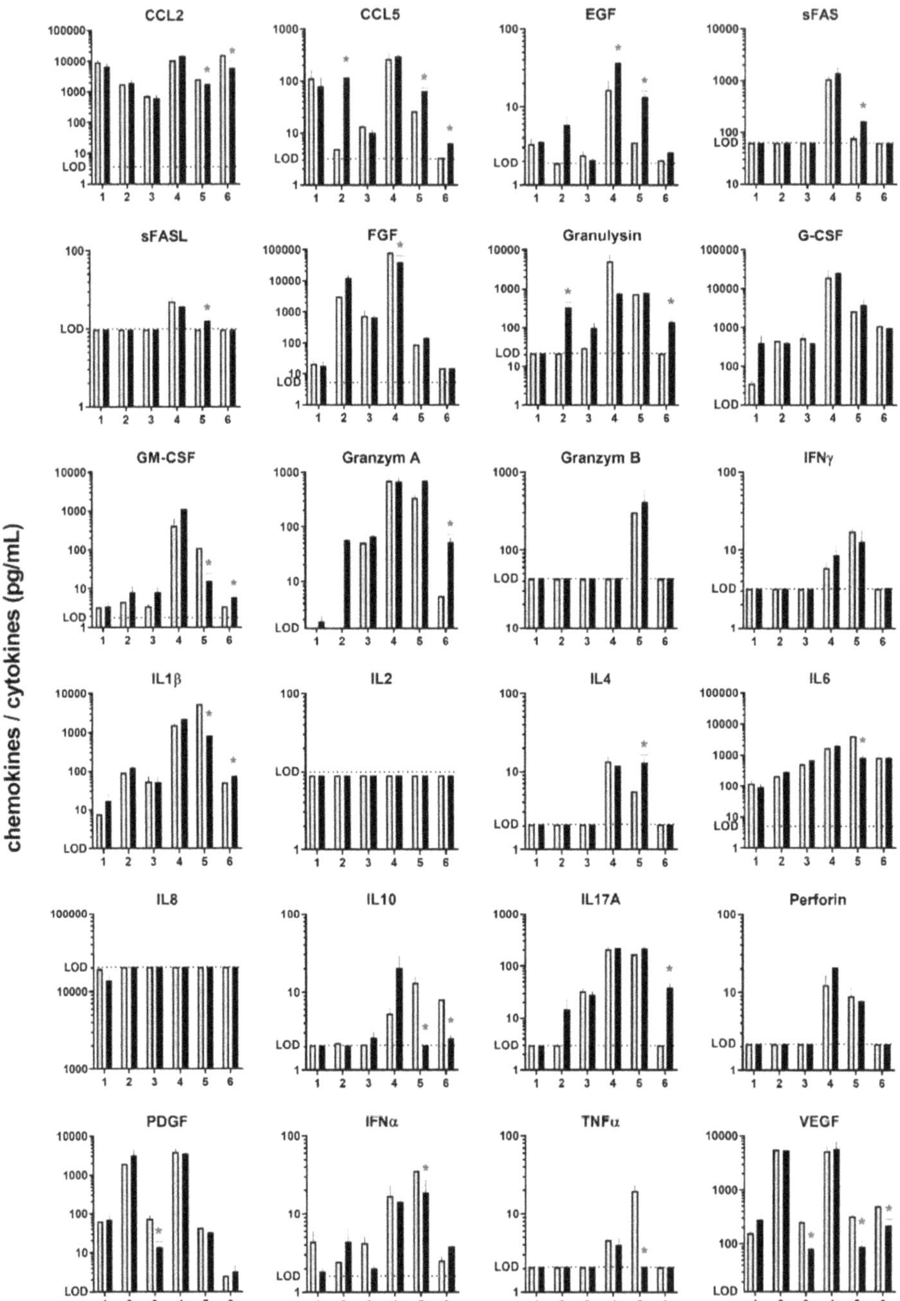

Figure 8.10. Chemokine/cytokine secretory profile of 24 targets measured in culture medium from cutaneous melanoma punch biopsy cultures 24 h after CAP treatment (dark bars) and the control group without treatment (grey bars). LOD: lower limit of detection, except for IL8 (upper limit of detection); red asterisks demark significant difference ($p < 0.05$) between CAP treatment and control groups for each patient (1–6). Reproduced from [64]. CC BY 4.0.

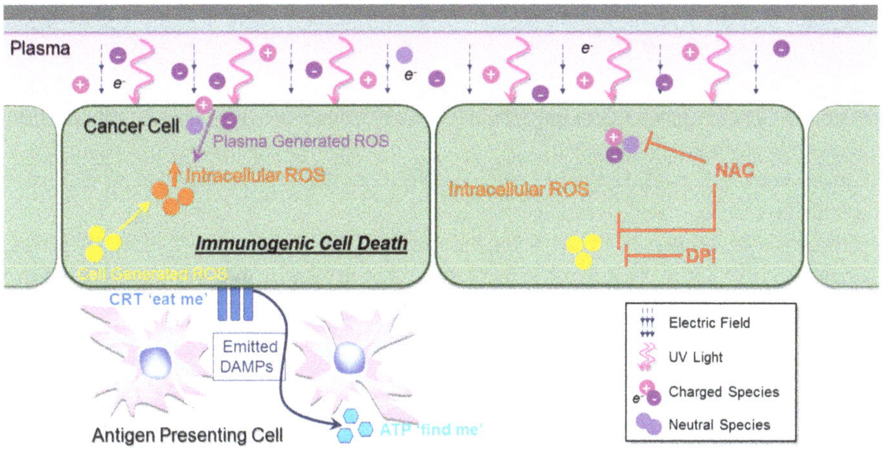

Figure 8.11. A CAP-based immunogenic cell death (ICD) model. Reactive species and other chemical factors may mainly contribute to eliciting ICD. CAP triggers the DAMP signals, such as ecto-CRT and ATP secretion. ATP acts as a 'find me' DAMP signal to recruit immune cells, and surface-exposed CRT serves as an eat me' signal. However, the physical factors' role is unknown [67]. CC By 4.0.

long-lived reactive species such as H_2O_2 *in vivo*. The tumorous tissue may be immersed in an H_2O_2-rich environment even though the CAP treatment does not directly generate H_2O_2 on the dry skin, which may explain the observed ROS rise in the CAP-treated mice [49]. In addition, H_2O_2 is a second messenger in lymphocyte activation [72, 73]. Micromole (μM) levels of H_2O_2 can rapidly induce the activation of the transcription factor NF-κB, early gene expression of interleukin-2 (IL-2), and the IL-2 receptor α chain [72]. If the tumorous tissue generates H_2O_2 due to the CAP treatment, such tissue may become a target for the immune system's attack.

Finally, physical factors in CAP may also play important roles in immune responses, particularly a CAP source was used to directly treat tumor tissues with or without a physical barrier such as skin [74]. The physical effect, particularly the EM effect in CAP, was not observed until the recent demonstration of a strong anti-cancer performance using a physically-based CAP treatment [75]. Physical effects in CAP have several novel features compared with chemical factors in CAP. Firstly, physical factors can penetrate a barrier to affect the target non-invasively, which matches the basic feature of cutaneous CAP treatment [38, 53]. More importantly, physical factors cause intense necrosis [53, 76, 77]. Necrosis will trigger inflammation and other immune responses *in vivo*, which explains the immune responses in many animal studies.

References

[1] Yan D, Malyavko A, Wang Q, Lin L and Sherman J H 2021 Cold atmospheric plasma cancer treatment, a critical review *Appl. Sci.* **11** 7757

[2] Mirpour S, Piroozmand S, Soleimani N, Jalali Faharani N, Ghomi H, Fotovat Eskandari H, Sharifi A M, Mirpour S, Eftekhari M and Nikkhah M 2016 Utilizing the micron sized non-

thermal atmospheric pressure plasma inside the animal body for the tumor treatment application *Sci. Rep.* **6** 29048
[3] Chen Z, Simonyan H, Cheng X, Gjika E, Lin L, Canady J, Sherman J H, Young C and Keidar M 2017 A novel micro cold atmospheric plasma device for glioblastoma both *in vitro* and *in vivo Cancers (Basel)* **9** 61
[4] Limanowski R, Yan D, Li L and Keidar M 2022 Preclinical cold atmospheric plasma cancer treatment *Cancers (Basel)* **14** 3461
[5] Vandamme M, Robert E, Pesnel S, Barbosa E, Dozias S, Sobilo J, Lerondel S, Pape A le and Pouvesle J M 2010 Antitumor effect of plasma treatment on U87 glioma xenografts: preliminary results *Plasma Processes Polym.* **7** 264–73
[6] Keidar M, Walk R, Shashurin A, Srinivasan P, Sandler A, Dasgupta S, Ravi R, Guerrero-Preston R and Trink B 2011 Cold plasma selectivity and the possibility of a paradigm shift in cancer therapy *Br. J. Cancer* **105** 1295–301
[7] Vandamme M, Robert E, Dozias S, Sobilo J, Lerondel S, Le Pape A and Pouvesle J-M 2011 Response of human glioma U87 xenografted on mice to non thermal plasma treatment *Plasma Med.* **1** 27–43
[8] Brullé L, Vandamme M, Riès D, Martel E, Robert E, Lerondel S, Trichet V, Richard S, Pouvesle J M and le Pape A 2012 Effects of a non thermal plasma treatment alone or in combination with gemcitabine in a MIA PaCa2-luc orthotopic pancreatic carcinoma odel *PLoS One* **7** e52653
[9] Vandamme M *et al* 2012 ROS implication in a new antitumor strategy based on non-thermal plasma *Int. J. Cancer* **130** 2185–94
[10] Walk R M, Snyder J A, Srinivasan P, Kirsch J, Diaz S O, Blanco F C, Shashurin A, Keidar M and Sandler A D 2013 Cold atmospheric plasma for the ablative treatment of neuroblastoma *J. Pediatr. Surg.* **48** 67–73
[11] Yajima I, Iida M, Kumasaka M Y, Omata Y, Ohgami N, Chang J, Ichihara S, Hori M and Kato M 2014 Non-equilibrium atmospheric pressure plasmas modulate cell cycle-related gene expressions in melanocytic tumors of RET-transgenic mice *Exp. Dermatol.* **23** 424–5
[12] Kang S U *et al* 2014 Nonthermal plasma induces head and neck cancer cell death: the potential involvement of mitogen-activated protein kinase-dependent mitochondrial reactive oxygen species *Cell Death Dis.* **5** e1056
[13] Chernets N, Kurpad D S, Alexeev V, Rodrigues D B and Freeman T A 2015 Reaction chemistry generated by nanosecond pulsed dielectric barrier discharge treatment is responsible for the tumor eradication in the B16 melanoma mouse model *Plasma Processes Polym.* **12** 1400–9
[14] Ikeda J I, Tsuruta Y, Nojima S, Sakakita H, Hori M and Ikehara Y 2015 Anti-cancer effects of nonequilibrium atmospheric pressure plasma on cancer-initiating cells in human endometrioid adenocarcinoma cells *Plasma Processes Polym.* **12** 1370–6
[15] Kaushik N K, Kaushik N, Yoo K C, Uddin N, Kim J S, Lee S J and Choi E H 2016 Low doses of PEG-coated gold nanoparticles sensitize solid tumors to cold plasma by blocking the PI3K/AKT-driven signaling axis to suppress cellular transformation by inhibiting growth and EMT *Biomaterials* **87** 118–30
[16] Binenbaum Y, Ben-David G, Gil Z, Slutsker Y Z, Ryzhkov M A, Felsteiner J, Krasik Y E and Cohen J T 2017 Cold atmospheric plasma, created at the tip of an elongated flexible capillary using low electric current, can slow the progression of melanoma *PLoS One* **12** e0169457

[17] Lin A G, Xiang B, Merlino D J, Baybutt T R, Sahu J, Fridman A, Snook A E and Miller V 2018 Non-thermal plasma induces immunogenic cell death *in vivo* in murine CT26 colorectal tumors *Oncoimmunology* **7** e1484978

[18] Yan D, Sherman J H and Keidar M 2017 Cold atmospheric plasma, a novel promising anti-cancer treatment modality *Oncotarget* **8** 15977

[19] Utsumi F, Kajiyama H, Nakamura K, Tanaka H, Mizuno M, Ishikawa K, Kondo H, Kano H, Hori M and Kikkawa F 2013 Effect of indirect nonequilibrium atmospheric pressure plasma on anti-proliferative activity against chronic chemo-resistant ovarian cancer cells *in vitro* and *in vivo PLoS One* **8** e81576

[20] Tanaka H, Nakamura K, Mizuno M, Ishikawa K and Takeda K 2016 Non-thermal atmospheric pressure plasma activates lactate in Ringer's solution for anti-tumor effects *Sci. Rep.* **6** 36282

[21] Nakamura K, Peng Y, Utsumi F, Tanaka H and Mizuno M 2017 Novel intraperitoneal treatment with non-thermal plasma-activated medium inhibits metastatic potential of ovarian cancer cells *Sci. Rep.* **7** 6085

[22] Takeda S *et al* 2017 Intraperitoneal administration of plasma-activated medium: proposal of a novel treatment option for peritoneal metastasis from gastric cancer *Ann. Surg. Oncol.* **24** 1188–94

[23] Liedtke K R, Bekeschus S, Kaeding A, Hackbarth C, Kuehn J-P, Heidecke C-D, von Bernstorff W, von Woedtke T and Partecke L I 2017 Non-thermal plasma-treated solution demonstrates antitumor activity against pancreatic cancer cells *in vitro* and *in vivo Sci. Rep.* **7** 8319

[24] Liedtke K, Freund E, Hermes M and Oswald S 2020 Gas plasma-conditioned Ringer's lactate enhances the cytotoxic activity of cisplatin and gemcitabine in pancreatic cancer *in vitro* and in ovo *Cancers* **12** 123

[25] Jiang L, Zheng H, Lyu Q, Hayashi S, Sato K and Sekido Y 2021 Lysosomal nitric oxide determines transition from autophagy to ferroptosis after exposure to plasma-activated Ringer's lactate *Redox Biol.* **43** 101989

[26] Yan D, Sherman J H and Keidar M 2018 The application of the cold atmospheric plasma-activated solutions in cancer treatment *Anti-cancer Agents Med. Chem.* **18** 769

[27] Rouven K, Freund E, Hackbarth C and Heidecke C 2018 A myeloid and lymphoid infiltrate in murine pancreatic tumors exposed to plasma-treated medium *Clin. Plasma Med.* **11** 10–7

[28] Takeda H, Bekeschus S, Yan D, Hori M, Keidar M and Laroussi M 2021 Plasma-treated solutions (PTS) in cancer therapy *Cancers* **13** 1737

[29] Harley J C, Suchowerska N and Mckenzie D R Cancer treatment with gas plasma and with gas plasma-activated liquid: positives, potentials and problems of clinical translation *Biophys. Rev.* **12** 989–1006

[30] Chen C Y, Cheng Y C and Cheng Y J 2018 Synergistic effects of plasma-activated medium and chemotherapeutic drugs in cancer treatment *J. Phys. D: Appl. Phys.* **51** 13LT01

[31] Yan D, Nourmohammadi N, Talbot A, Sherman J H and Keidar M 2016 The strong anti-glioblastoma capacity of the plasma-stimulated lysine-rich medium *J. Phys. D: Appl. Phys.* **49** 274001

[32] Kaushik N, Lee S J, Choi T G, Baik K Y, Uhm H S, Kim C H, Kaushik N K and Choi E H 2015 Non-thermal plasma with 2-deoxy-D-glucose synergistically induces cell death by targeting glycolysis in blood cancer cells *Sci. Rep.* **5** 8726

[33] Xu D, Cui Q, Xu Y, Wang B, Tian M, Li Q, Liu Z, Liu D, Chen H and Kong M G 2018 Systemic study on the safety of immuno-deficient nude mice treated by atmospheric plasma-activated water *Plasma Sci. Technol.* **20** 44003

[34] Nastasa V, Pasca A S, Malancus R N, Bostanaru A C, Ailincai L I, Ursu E L, Vasiliu A L, Minea B, Hnatiuc E and Mares M 2021 Toxicity assessment of long-term exposure to non-thermal plasma activated water in mice *Int. J. Mol. Sci.* **22** 11534

[35] Zhang H *et al* 2021 Plasma-activated thermosensitive biogel as an exogenous ROS carrier for post-surgical treatment of cancer *Biomaterials* **276** 121057

[36] Mizuno K, Yonetamari K, Shirakawa Y, Akiyama T and Ono R 2017 Anti-tumor immune response induced by nanosecond pulsed streamer discharge in mice *J. Phys. D: Appl. Phys.* **50** 12LT01

[37] Jinno R, Komuro A, Yanai H and Ono R 2022 Antitumor abscopal effects in mice induced by normal tissue irradiation using pulsed streamer discharge plasma *J. Phys. D: Appl. Phys.* **55** 17LT01

[38] Yan D, Wang Q, Yao X and Malyavko A 2021 Anti-melanoma capability of contactless cold atmospheric plasma treatment *Int. J. Mol. Sci.* **22** 11728

[39] Xu W, Yan D, Sun J, Chen J, Yao X, Sherman J H and Keidar M 2020 The activation of cancer cells by a nanosecond-pulsed magnetic field generator *J. Phys. D: Appl. Phys.* **53** 125401

[40] Yan D, Lin L, Xu W, Nourmohammadi N, Sherman J and Keidar M 2019 Cold plasma-based control of the activation of pancreatic adenocarcinoma cells *J. Phys. D: Appl. Phys.* **52** 445202

[41] Yan D, Xu W, Yao X, Lin L, Sherman J H and Keidar M 2018 The cell activation phenomena in the cold atmospheric plasma cancer treatment *Sci. Rep.* **8** 15418

[42] Eramo A, Ricci-Vitiani L, Zeuner A, Pallini R, Lotti F, Sette G, Pilozzi E, Larocca L M, Peschle C and de Maria R 2006 Chemotherapy resistance of glioblastoma stem cells *Cell Death Differ.* **13** 1238–41

[43] Lee S Y 2016 Temozolomide resistance in glioblastoma multiforme *Genes Dis.* **3** 198–210

[44] Yao X, Lin L, Soni V, Gjika E, Sherman J H, Yan D and Keidar M 2020 Sensitization of glioblastoma cells to temozolomide by a helium gas discharge tube *Phys. Plasma* **27** 114502

[45] Soni V, Adhikari M, Simonyan H, Lin L, Sherman J H, Young C N and Keidar M 2021 *in vitro* and *in vivo* enhancement of temozolomide effect in human glioblastoma by non-invasive application of cold atmospheric plasma *Cancers (Basel)* **13** 4485

[46] Yao X, Yan D, Lin L, Sherman J H, Peters K B, Keir S T and Keidar M 2022 Cold plasma discharge tube enhances antitumoral efficacy of temozolomide *ACS Appl. Bio Mater.* **5** 1610–23

[47] Graves D B 2014 Oxy-nitroso shielding burst model of cold atmospheric plasma therapeutics *Clin. Plasma Med.* **2** 38–49

[48] Graves D B 2014 Reactive species from cold atmospheric plasma: implications for cancer therapy *Plasma Processes Polym.* **11** 1120–7

[49] Szili E J *et al* 2017 Modelling the helium plasma jet delivery of reactive species into a 3D cancer tumour *Plasma Sources Sci. Technol.* **27** 014001

[50] Lu X, Keidar M, Laroussi M, Choi E, Szili E J and Ostrikov K 2019 Transcutaneous plasma stress: from soft-matter models to living tissues *Mater. Sci. Eng. R. Rep.* **138** 36–59

[51] Partecke L I *et al* 2012 Tissue tolerable plasma (TTP) induces apoptosis in pancreatic cancer cells *in vitro* and *in vivo* *BMC Cancer* **12** 473

[52] Wenzel T, Carvajal Berrio D A, Daum R, Reisenauer C, Weltmann K D, Wallwiener D, Brucker S Y, Schenke-Layland K, Brauchle E M and Weiss M 2019 Molecular effects and tissue penetration depth of physical plasma in human mucosa analyzed by contact- and marker-independent Raman microspectroscopy *ACS Appl. Mater. Interfaces* **11** 42885–95

[53] Yan D, Wang Q, Adhikari M, Malyavko A, Lin L, Zolotukhin D B, Yao X, Kirschner M, Sherman J H and Keidar M 2020 A physically triggered cell death via transbarrier cold atmospheric plasma cancer treatment *ACS Appl. Mater. Interfaces* **12** 34548–63

[54] Jun-Seok O, Szili E J, Gaur N, Hong S H, Furuta H, Kurita H, Mizuno A, Hatta A and Short R D 2016 How to assess the plasma delivery of RONS into tissue fluid and tissue *J. Phys. D: Appl. Phys.* **49** 304005

[55] Szili E J, Hong S, Oh J, Gaur N and Short R D Tracking the penetration of plasma reactive species in tissue models *Trends Biotechnol.* **36** 594–602

[56] Gaur N, Szili E J, Oh J-S, Hong S-H, Michelmore A, Graves D B, Hatta A and Short R D 2015 Combined effect of protein and oxygen on reactive oxygen and nitrogen species in the plasma treatment of tissue *Appl. Phys. Lett.* **107** 103703

[57] Liu X, Gan L, Ma M, Zhang S and Liu J 2018 A comparative study on the transdermal penetration effect of gaseous and aqueous plasma reactive species *J. Phys. D: Appl. Phys.* **51** 075401

[58] Omran A V, Busco G, Ridou L, Dozias S, Grillon C, Pouvesle J-M and Robert E 2020 Cold atmospheric single plasma jet for RONS delivery on large biological surfaces *Plasma Sources Sci. Technol.* **29** 105002

[59] Ki S H, Masur K, Baik K Y and Choi E H 2021 UV absorption spectroscopy for the diffusion of plasma-generated reactive species through a skin model *Appl. Sci. (Switzerland)* **11** 7958

[60] Jiang C, Oshin E A, Guo S, Scott M, Li X, Mangiamele C and Heller R 2021 Synergistic effects of an atmospheric-pressure plasma jet and pulsed electric field on cells and skin *IEEE Trans. Plasma Sci.* **49** 3317–24

[61] Hashimoto S *et al* 2023 Understanding the role of plasma bullet currents in heating skin to mitigate risks of thermal damage caused by low-temperature atmospheric-pressure plasma jets *Plasma* **6** 103–14

[62] Miller V, Lin A, Fridman G, Dobrynin D and Fridman A 2014 Plasma stimulation of migration of macrophages *Plasma Processes Polym.* **11** 1193–7

[63] Cheng F, Yan D, Chen J, Wang Z, Horkowitz A, Keidar M and Sotomayor E M 2022 Enhancing innate and adaptive immune systems by cold atmospheric plasma (CAP) and its antitumor immunity arXiv:2201.12737

[64] Bekeschus S, Moritz J, Helfrich I, Boeckmann L, Weltmann K D, Emmert S, Metelmann H R, Stoffels I and von Woedtke T 2020 Ex vivo exposure of human melanoma tissue to cold physical plasma elicits apoptosis and modulates inflammation *Appl. Sci. (Switzerland)* **10** 1971

[65] Miller V, Lin A and Fridman A 2016 Why target immune cells for plasma treatment of cancer *Plasma Chem. Plasma Process.* **36** 259–68

[66] Lin A *et al* 2015 Uniform nanosecond pulsed dielectric barrier discharge plasma enhances anti-tumor effects by induction of immunogenic cell death in tumors and stimulation of macrophages *Plasma Processes Polym.* **12** 1392–9

[67] Lin A, Truong B, Patel S, Kaushik N, Choi E H, Fridman G, Fridman A and Miller V 2017 Nanosecond-pulsed DBD plasma-generated reactive oxygen species trigger immunogenic cell death in A549 lung carcinoma cells through intracellular oxidative stress *Int. J. Mol. Sci.* **18** 966

[68] Bekeschus S, Mueller A, Miller V, Gaipl U and Weltmann K 2017 Physical plasma elicits immunogenic cancer cell death and mitochondrial singlet oxygen *IEEE Trans. Radiat. Plasma Med. Sci.* **2** 138–46

[69] Mohamed H, Esposito R A, Kutzler M A, Wigdahl B, Krebs F C and Miller V 2020 Nonthermal plasma as part of a novel strategy for vaccination *Plasma Processes Polym.* **17** 2000051

[70] Khalili M, Daniels L, Lin A, Krebs F C, Snook A E, Bekeschus S, Bowne W B and Miller V 2019 Non-thermal plasma-induced immunogenic cell death in cancer *J. Phys. D: Appl. Phys.* **52** 423001

[71] Yan D, Cui H, Zhu W, Talbot A, Zhang L G, Sherman J H and Keidar M 2017 The strong cell-based hydrogen peroxide generation triggered by cold atmospheric plasma *Sci. Rep.* **7** 10831

[72] Los M *et al* 1995 Hydrogen peroxide as a potent activator of T-lymphocyte functions *Eur. J. Immunol.* **25** 159–65

[73] Reth M 2002 Hydrogen peroxide as second messenger in lymphocyte activation *Nat. Immunol.* **3** 1129–34

[74] Graves D B 2018 Lessons from Tesla for plasma medicine *IEEE Trans. Radiat. Plasma Med. Sci.* **2** 594–607

[75] Yan D, Malyavko A, Wang Q, Ostrikov K K, Sherman J H and Keidar M 2021 Multi-modal biological destruction by cold atmospheric plasma: capability and mechanism *Biomedicines* **9** 1259

[76] Wang Q, Malyavko A, Yan D, Lamanna O K, Hsieh M H, Sherman J and Keidar M 2020 A comparative study of cold atmospheric plasma treatment, chemical versus physical strategy *J. Phys. D: Appl. Phys.* **54** 095207

[77] Yan D, Wang Q, Malyavko A, Zolotukhin D B, Adhikari M and Keidar M 2020 The anti-glioblastoma effect of cold atmospheric plasma treatment: physical pathway v.s. chemical pathway *Sci. Rep.* **10** 11788

Chapter 9

Clinical perspectives

The clinical application of cold atmospheric plasma (CAP) as an anti-tumor modality is the ultimate goal in plasma medicine. To date, CAP-based clinical tests in cancer therapy are still relatively rare. Various organizations have performed clinical tests where CAP treatment trials were incorporated into patient treatment plans. A few examples are illustrated here, providing critical clues and inspiration to guide the use of CAP in therapeutic advances.

9.1 Clinical studies

Overcoming technical barriers of using CAP and finding an optimized and practical strategy to explore the clinical anti-cancer potential of CAP technology are complex missions for plasma medicine that lie ahead. However, the clinical use of CAP in cancer treatment is still rare. In 2015, a private company, US Medical Innovation (USMI), conducted a clinical trial on stage 4 metastatic colon cancer at Baton Rouge General Medical Center in Baton Rouge, Louisiana, USA. CAP treatment was performed on the post-surgical tissue to inhibit potential residual tumor tissue's growth after a removal surgery, and no relapse or progression of cancer occurred in patients. This trial also demonstrated the safety of CAP treatment [1]. In Germany, CAP trials were performed on 12 patients with advanced squamous cell carcinoma of the head and neck [2]. CAP was also used to decontaminate infected cancer ulcerations during this trial. Generally, this trial generated positive patient effects, including a decreased request for pain medication and a reduction of typical fetid odor and microbial load [2]. In some cases, superficial partial remission of tumors and even wound healing of the infected ulcerations have been observed as well [2].

In 2017, USMI used Canady Helios Cold Plasma and Hybrid Plasma Scalpels in clinical liver resection to remove and selectively kill liver tumor cells [1]. In the

doi:10.1088/978-0-7503-5537-7ch9

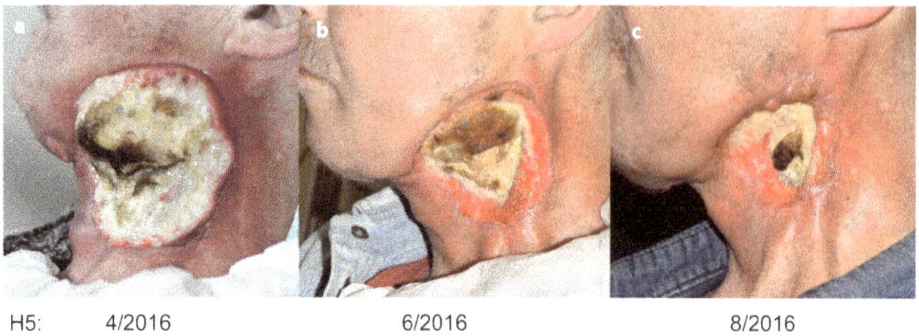

Figure 9.1. Clinical effect of CAP trials on a patient (H5) with locally advanced head and neck cancer. Reproduced from [3], copyright (2018), with permission from Elsevier.

same year, another impactful treatment was performed by Metelmann *et al* [3]. Their trial enrolled six patients with local advanced (pT4) squamous cell carcinoma of the oropharynx with open infected ulcerations. Six patients were treated by an atmospheric pressure plasma jet (APPJ) in a cycle of three single applications within a week, each followed by an intermittence of another week. As shown in figure 9.1, the trial noticeably improved the therapeutic efficacy of this locally advanced head and neck cancer. CAP treatment caused a reduction in odor and pain medication requirement, which improved the social function of patients and generated a positive emotional effect. Partial remission in two patients for at least nine months has been observed. A moderate amount of apoptotic tumor cells and a desmoplastic reaction of the connective tissue was observed in the incisional biopsies. CAP was also used to treat oral cancer with very mild side effects, such as bad taste during and after treatment and minimal discomfort during treatment. None of these side effects were life-threatening; most were transient, with full recovery of normalcy [4]. Besides, these clinical trials also strongly indicated that the widely observed wound healing capability in CAP treatment might play a critical supporting role in cancer therapy [5].

In 2022, USMI presented the results of a two-year follow-up phase I clinical trial of Canady Helios Cold Atmospheric Plasma (CHCP) treatment for patients with advanced stage IV metastatic and recurrent solid tumors at the Biannual Conference of the Israeli Society of Surgical Oncology. Twenty patients were recruited from the US and Israel. Patients received intra-operative CHCP treatment at the operative site after removing tumors [6]. For this project, the primary endpoint was safety, and the secondary point was to demonstrate ablation and slowing down tumor growth in cancer patients without damaging surrounding normal tissues [7]. Currently, the primary endpoint is satisfying and promising. Together, the existing clinical trials suggest that CAP treatment can be a powerful supplemental tool to improve the current surgery and even chemotherapy efficacy.

Current clinical tests have a common feature: bulk plasma will directly touch tumorous tissues without the cover of an aqueous solution layer at a macro scale. Such working conditions are essentially different from the experimental conditions that are always used in *in vitro* studies. The absence of an apparent aqueous layer probably makes the physical effects, such as the EM effect and thermal effect of CAP, play some unignorable roles in the clinical application [6].

9.2 Outlook

Here, potential routes of CAP-based cancer therapy can be proposed (figure 9.2). For subcutaneous or cutaneous tumors, a direct treatment strategy, such as just above or near the skin of the tumor site is recommended. According to the observation from physically-based CAP treatment *in vitro*, we speculate that even an air gap such as several mm between bulk plasma and skin will not fully weaken CAP's therapeutic efficacy [8]. In contrast, the treatment efficacy may be largely inhibited in some cases when bulk plasma completely touches the skin [9]. Regarding intraperitoneal tumors or tumors in deeper tissues, just treating above the skin is a questioning strategy based on the current studies. The effective penetration depth of physical factors from CAP is still unknown in plasma medicine. In these cases, PAS may be used to treat the tumor by administering intraperitoneal injections. Furthermore, CAP sources may be synergistically used with endoscopic devices to treat deep tissue tumors by minimally invasive surgery [10]. More attractively, CAP can also be used as an adjunct to current surgical approaches. Bulk plasma or PAS can come in direct contact with the exposed tumor tissue or the remnants of tumor tissue during surgery [11]. This strategy was used in some clinical trials performed by USMI [7]. Finally, CAP may be used to enhance chemotherapeutic efficacy or selectivity by its sensitization power revealed in recent studies, particularly for deep tumors like glioblastoma in the skull. Discharge tubes may be used to selectively enhance the TMZ's efficacy on glioblastoma by just treating above the skin of head, which is a promising candidate for a plasma-based non-invasive modality [12].

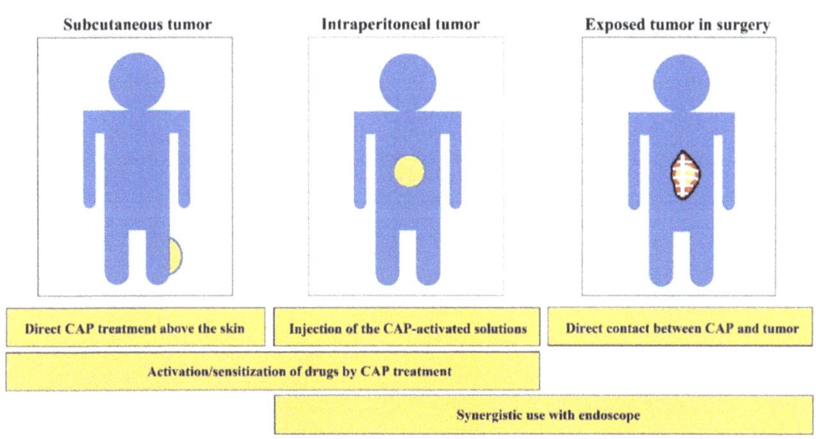

Figure 9.2. Schematic illustration of possible ways to use CAP in clinical cancer. Reproduced from [8]. CC BY 4.0.

References

[1] Canady J, Trink B, Sherman J, Keidar M, Maho T, Damany X, Dozias S, Pouvesle J and Robert E 2018 Cold atmospheric plasma (CAP) combined with chemo-radiation and cytoreductive surgery: the first clinical experience for stage IV metastatic colon cancer *Comprehensive Clinical Plasma Medicine: Cold Physical Plasma for Medical Application* ed Metelmann H R, Von Woedtke T and Weltmann K D (Cham: Springer-Nature) pp 163–83

[2] Metelmann H-R *et al* 2015 Head and neck cancer treatment and physical plasma *Clin. Plasma Med.* **3** 17–23

[3] Metelmann H R *et al* 2018 Clinical experience with cold plasma in the treatment of locally advanced head and neck cancer *Clin. Plasma Med.* **9** 6–13

[4] Schuster M, Rutkowski R, Hauschild A, Shojaei R K, von Woedtke T, Rana A, Bauer G, Metelmann P and Seebauer C 2018 Side effects in cold plasma treatment of advanced oral cancer—clinical data and biological interpretation *Clin. Plasma Med.* **10** 9–15

[5] Bekeschus S, von Woedtke T, Emmert S and Schmidt A 2021 Medical gas plasma-stimulated wound healing: evidence and mechanisms *Redox Biol.* **46** 102116

[6] Limanowski R, Yan D, Li L and Keidar M 2022 Preclinical cold atmospheric plasma cancer treatment *Cancers (Basel)* **14** 3461

[7] Canady J *et al* 2023 The first cold atmospheric plasma phase I clinical trial for the treatment of advanced solid tumors: a novel 4th treatment arm for cancer *J. Clin. Oncol.* **41** e14596

[8] Yan D, Malyavko A, Wang Q, Lin L and Sherman J H 2021 Cold atmospheric plasma cancer treatment, a critical review *Appl. Sci.* **11** 7757

[9] Yan D, Wang Q, Adhikari M, Malyavko A, Lin L, Zolotukhin D B, Yao X, Kirschner M, Sherman J H and Keidar M 2020 A physically triggered cell death via transbarrier cold atmospheric plasma cancer treatment *ACS Appl. Mater. Interfaces* **12** 34548–63

[10] Robert E *et al* 2013 Perspectives of endoscopic plasma applications *Clin. Plasma Med.* **1** 8–16

[11] Zhang H *et al* 2021 Plasma-activated thermosensitive biogel as an exogenous ROS carrier for post-surgical treatment of cancer *Biomaterials* **276** 121057

[12] Yao X, Yan D, Lin L, Sherman J H, Peters K B, Keir S T and Keidar M 2022 Cold plasma discharge tube enhances antitumoral efficacy of temozolomide *ACS Appl. Bio Mater.* **5** 1610–23

IOP Publishing

Cold Atmospheric Plasma-based Cancer Therapy
(Second Edition)

Michael Keidar, Dayun Yan and Li Lin

Chapter 10

Physics and technology of the adaptive plasma therapy

In chapter 10, we describe recently proposed new approaches for plasma therapy based on plasma adaptation. The adaptive plasma approach is based on the ability to read the cellular response to cold atmospheric plasma (CAP) in real time and modify the composition and power of the plasma via a feedback mechanism. In addition, plasma self-adaptation might be feasible due to self-organization and pattern formation when plasma interacts with targets. Both strategies will be described in this chapter.

10.1 Plasma self-organization phenomena

Nature consists of various types of patterns and coherent structures ranging from microscopic organisms and colonies to snowflakes, crystals, as well as structures seen throughout the Galaxy. To explain the various phenomena associated with pattern formation, W Ross Ashby proposed the so-called 'principle of self-organization' [1]. It has been suggested that a dynamic system, independent of its type or composition, always evolves towards a state of equilibrium or an attractor. Such tendencies reduce uncertainty about the system's state, reducing the system's statistical entropy. Self-organization is a complex and fascinating phenomenon commonly observed in both natural and technological settings within diverse varieties of physics, chemistry, and biology [2]. Self-organization in plasmas can be described as a spontaneous transition from a homogeneous, stable state to a regular pattern in a spatially extended system or a transition between different patterns [3]. Different types of self-organization phenomena have been reported in a wide range of plasmas, such as dielectric barrier discharge [4], high-frequency discharge [5], gas flow stabilized discharges [6], resistively stabilized discharges [7], and discharges with liquid electrodes [8]. A recent review analyzed various aspects of spatial and temporal

doi:10.1088/978-0-7503-5537-7ch10

pattern formation resulting from self-organization [9]. In many situations, pattern formation results from a nonlinear interaction of the plasmas with the electrode surfaces.

Recent experiments demonstrated the effect of self-organized plasma on the production of chemical species and cancer treatment. Figure 10.1 shows optical photographs of various self-organized stratified interface patterns with micro-discharges [10]. Complex shapes consisting of radial and confocal lines of different densities were produced. Some elements of symmetry (both radial and axial) can be sometimes observed. The current–voltage characteristics and optical photograph discharge patterns above the DI water corresponding to the specific current/voltage conditions are shown in figure 10.1(d). The whole current–voltage characteristic can be divided into four specific stages. At stage I (current not exceeding 7 mA), the discharge voltage and current are low, and the discharge pattern represents a single filament (i.e., contracted glow discharge) following the initial corona discharge. As the discharge voltage increases to a certain degree (stage II), the temperature of the tungsten cathode increases, resulting in drastically enhanced heat radiation, as seen in figure 10.1(c). At stage III, the discharge enters an unstable state where it alternates between two (II and IV) stages featuring strong heat radiation (stage II)

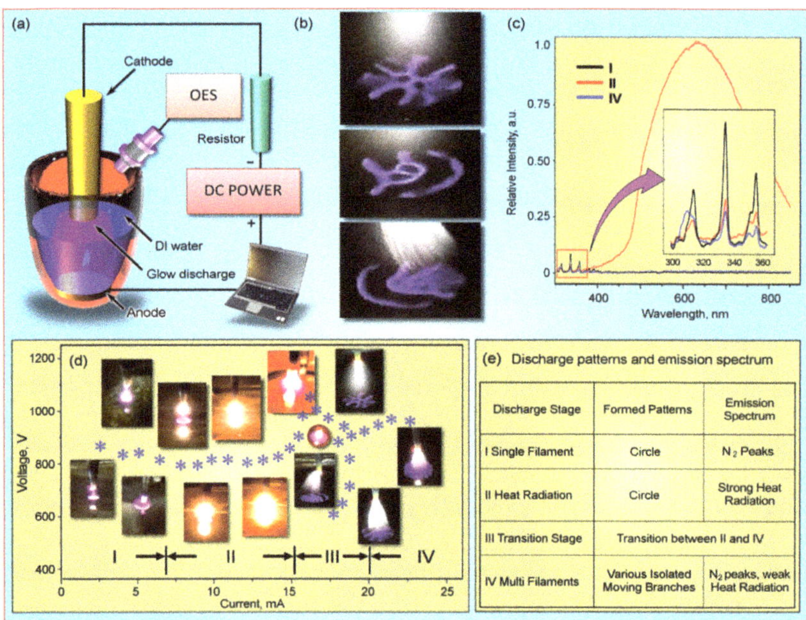

Figure 10.1. Discharge self-organization. (a) Schematic representation of the atmospheric pressure glow micro-discharge setup. (b) Optical photographs of the discharge patterns above DI water. The self-organized patterns have a complex structure strongly depending on the voltage-current conditions. (c) Optical emission spectrum by the atmospheric pressure glow discharge above DI water. (d) Current–voltage dependence of the system with optical photographs of the self-organized stratified interface patterns. Four discharge stages could be clearly determined. (e) The relationship between discharge stages, self-organized patterns, and optical emission. Reproduced from [10]. CC BY 4.0.

and the multi-filament pattern (stage IV). A plausible reason for this behavior could be that the high-temperature results in stronger thermionic rather than secondary electron emission from a comparatively cold electrode. Thus, the discharge flips from the multi-filament and heat radiation-supported stage where the thermionic emission is efficient enough to maintain the increased discharge current. At stage IV, the discharge stabilizes at the multi-filament stage and stretches out to a large number of discharge filaments at the liquid media surface. Various self-organized stratified interface patterns are formed above the liquid media surface, as shown in figures 10.1(b) and (c).

Self-organized plasma may have important applications in biomedical engineering. The formation of self-organized patterns in plasmas because of instability and bifurcation could lead to the formation of coherent structures. These coherent structures might modulate plasma chemistry and composition, including reactive species, electric fields, and charged particles. As a first step in understanding the effect of self-organization on cell treatment, we should consider an adaptive system that utilizes the feedback mechanism. An example of this phenomenon is shown in figure 10.2(a). Glioblastoma cells are generally more resistant to CAP treatment

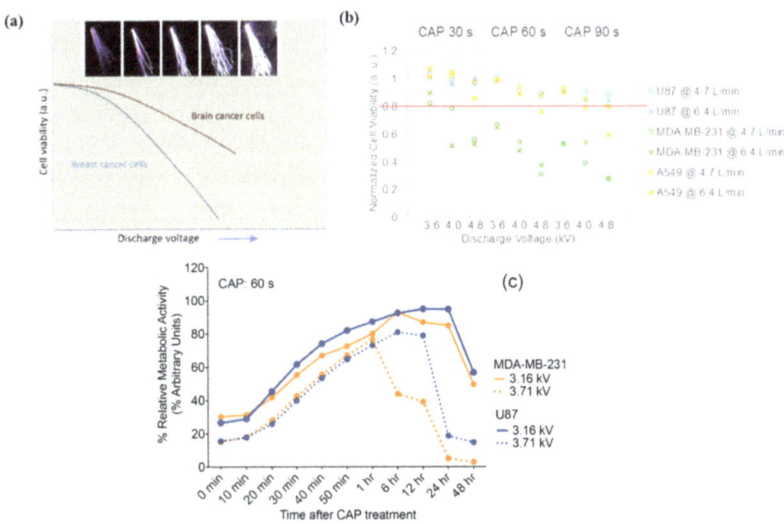

Figure 10.2. The cytotoxicity of CAP over cancer cell lines can be modulated by plasma parameters. (a) The effect of discharge voltage on the cytotoxicity of CAP over glioblastoma cells (U87MG) and breast cancer cells (MDA-MB-231). Insert: plasma self-organization patterns above the water under different discharge voltages. Reprinted from [11], copyright (2018), with permission from Elsevier. (b) Discharge voltage and flow rate are two important factors for CAP control. Compared with the gas flow rate, the CAP discharge voltage is a more important factor affecting the death rate of glioblastoma cells (U87MG), breast cancer cells (MDA-MB-231), as well as lung cancer cells (A549). Target cell viability is marked as 0.8, i.e. 80% of cells are killed after CAP treatment [12]. (c) Breast cancer cells (MDA-MB-231) and glioblastoma cells (U87) were treated with a CAP output voltage of 3.16 and 3.71 kV for 60 s. The RealTime-Glo Cell Viability Assay (Promega) performed with the continuous read method indicates differences in cell viability among the cell lines with dependency on the CAP discharge voltage [12]. (b) and (c) Reprinted with permission from [12], copyright (2018) American Chemical Society.

than breast cancer cells. An increase in the discharge voltage can lead to stronger effects of cold plasma treatment. In addition, one can see that the plasma composition can also be affected by other discharge conditions, such as flow rate and power. Consequently, to reach equivalent levels of a specifically targeted cell viability, different cancer cell lines each have a unique combination of helium flow rate and discharge voltage (figure 10.2(b)). Thus, the plasma–cell interaction outcome can be affected by discharge conditions such as gas flow rate and voltage. It is suggested that an adaptive plasma technique can be based on the ability to change plasma composition and key plasma parameters (such as density and temperature) on demand, dependent on specific applications and needs. Moreover, continuous recording of the cellular response to CAP is possible, as demonstrated in figure 10.2(c). One can see that the immediate response can differentiate between discharge voltage and cell type difference.

Figure 10.2(c) shows the effect of the discharge voltage on cancer cell viability utilizing the RealTime-Glo MT Cell Viability Assay. This assay enables continuous monitoring of the cellular response via a repeated reading of the testing conditions over 48 h. The RealTime-Glo assay involves the addition of a proprietary luciferase and a pro-substrate to the culture medium before CAP exposure. Live cells unaffected by CAP treatment—facilitate the conversion of the pro-substrate into a new substrate, which then diffuses into the culture media. The newly generated substrate is consumed by luciferase to produce a luminescent signal. The signal from RealTime-Glo MT correlates with the number of viable cells after treatment since dead cells are unable to reduce the pro-substrate and therefore did not produce a signal. The percentage of viability of cells is determined with respect to the untreated control condition. More specifically, the percentage of viability is calculated by dividing luminescent readings of the CAP-treated conditions with those of the untreated control and multiplying by 100%. Therefore, the cell viability of the untreated control is assumed as 100% and is indicated by the dotted line.

To that end, the viability of breast cancer and glioblastoma cells was quantified for up to 48 h following CAP exposures of 30, 60, 90, and 180 s at 3.16 kV, as shown in figure 10.3. A greater than 50% reduction in cell viability was recorded immediately (0 min) after treatment in both cell lines for most CAP exposure times except for 180 s treatment in breast cancer cells. Data showed that 30–90 s CAP treatments lead to more significant reductions in viability than 180 s CAP exposure. This unexpected outcome may result from CAP-generated reactive species needing time to fully become incorporated in cell culture media and come in contact with cancer cells. In fact, according to the data, it was not until 20 min after CAP treatment that the differences in cell viability correlated with treatment time duration. The decrease in cell viability observed immediately after treatment was followed by a viability increase which lasted up to 6 h. Cell viability steadily decreased from 6 to 48 h as a function of exposure time for most CAP conditions. The delayed effect of plasma in initiating cell death may depend on the phase of the cell cycle.

Despite exhibiting a similar cell viability trend in the initial hour after treatment, glioblastoma cell response differed significantly from breast cancer cells between 6

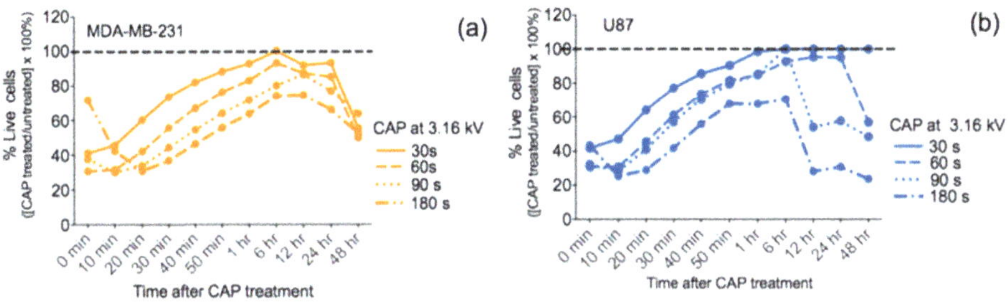

Figure 10.3. CAP treatment induces a significant reduction in cell viability over time as a function of exposure time. MDA-MB-231 breast cancer cells (a) and U87 glioblastoma (b) cells were treated with a CAP output voltage of 3.16 kV for 30, 60, 90 and 180 s. Immediately following treatment, RealTime-Glo MT Cell Viability Assay (Promega) was performed with a continuous read method to evaluate the response of each cell line in real-time. The data were normalized to the untreated control recorded for each time point. The % cell viability of the untreated control was assumed as 100% as indicated by the dotted line. Reprinted with permission from [12], copyright (2018) American Chemical Society.

and 48 h. At 24 and 48 h, the shortest CAP exposure of 30 s ultimately amounted to no substantial reduction of glioblastoma cell viability, while the viability of breast cancer cells decreased by 10%–30%. Plasma treatments of 60, 90, and 180 s had a more significant effect on cell viability. These results attest that different cell lines responded differently to treatment.

According to the 24 h data collected from the 60 s treatment condition, glioblastoma and breast cancer cell viability were reported at 95% and 85%, respectively. The variation in response to treatment between the two cancer cell lines was noted for all CAP exposure times. The data provided evidence that the RealTime-Glo assay could detect the cell response to plasma treatment in real time with results varying per cell line and in a CAP-exposure time-dependent manner.

Figure 10.2(c) presents the cell response pattern corresponding to the 60-s treatment. The data emphasize the variation in cell response resulting from an increase in discharge voltage from 3.16 to 3.71 kV. Breast cancer data revealed a statistically significant difference in cell viability between 3.16 and 3.71 kV conditions in the 6–48 h time range. The earliest detectable variance in cell response observed 6 h post-treatment corresponded to a 50% final difference in cell viability between the two voltage conditions. The first significant difference in cell response was observed in glioblastoma cells as early as 20 min after treatment. The discharge voltage effect became more pronounced at 24 h, where a 0.55 kV increase in voltage resulted in a 70% reduction in viability. Overall, real-time results demonstrate the important role of CAP discharge voltage in influencing cell response. These results also suggest that instantaneous cellular response might be used for adapting plasma conditions to achieve the desired outcome.

To enable a platform that can operate on self-adaptive plasma features, one needs to understand the mechanism of plasma's interaction with cells, particularly the effect of cells on the cold atmospheric plasma.

10.2 Adaptive plasma devices

In this chapter, we briefly describe new ideas of adaptive CAP as shown schematically in figure 10.4. The concept is based on the ability to read the cellular response to CAP in real-time and to modify the plasma composition and power via a feedback mechanism [11, 12]. This will allow for the real-time modification of the plasma conditions to optimize the cold plasma effect on cancer and normal cells. In addition, due to the distinct effect of cold plasma on normal cells and cancer cells, such a system leads to enhanced plasma selectivity by controlling the generation of reactive species or the potential interaction between CAP and cells [13, 14]. Figure 10.4(b) shows schematically how the plasma interaction with cells can lead to plasma transition between different plasma discharge modes, thus allowing plasma self-adaptation. The uniqueness of plasma is its ability to change its composition and key parameters on demand, dependent on specific requirements [11].

It is possible to devise an adaptive plasma therapeutic system to adjust its interaction with cells and tissues by responding to the boundary conditions at the plasma–cell interface. In that respect, the chemistry of discharge modes requires a detailed study. In addition, new plasma sources with many operation modes with bifurcation points should be developed. To implement an adaptive plasma system, an efficient multi-parametric feedback system based on cellular responses is required. The nature of the plasma adaptive therapeutic system through a real-time cellular and tissue response can lead to optimal and selective treatment. An example of the feedback mechanism will be described in the next section.

Such an approach can ultimately lead to personalized plasma-based treatment of cancer and possibly other diseases. Personalized medicine is currently being

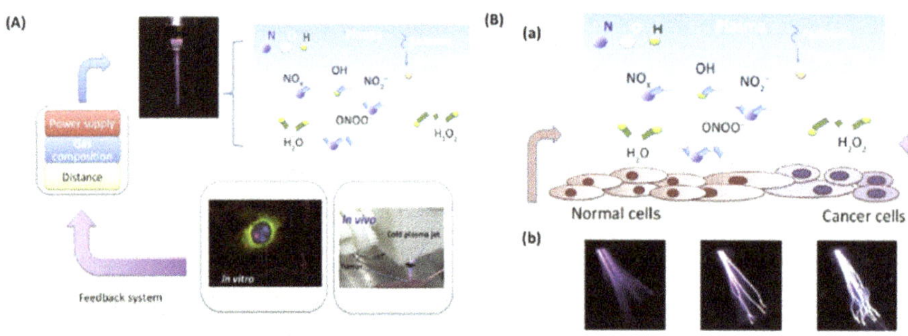

Figure 10.4. Adaptive plasma concept. (a) Plasma adaptation. CAP *in vivo* or *in vitro* treatment will trigger differing responses in cancer and normal cells. A plasma system will measure, record, and analyze the signal based on these responses. The chemical composition will be modulated through changes in the power supply, gas composition, and the distance between the plasma source and the cells through the feedback system. (b) Plasma self-adaptation-based self-organization and pattern formation. Cancer and normal cells exhibit significant differences in membrane potential, which can be used as a signal to trigger plasma self-organization. As a result, cold plasma is affected differently by normal and cancer cells. The distinct response of the plasma to cells and tissue leads to plasma mode transition. Reprinted from [11], copyright (2018), with permission from Elsevier.

investigated in cancer research therapy, and an adaptive plasma-based system would fit well into this multi-disciplinary treatment approach. The main idea of personalized medicine is to customize treatment to each patient's needs based on their genetic makeup. The plasma application might lead to specific unique responses dependent on the person's genetic makeup. Thus, due to genome specifics, the same combination of reactive species and the electric field applied to the body during the plasma treatment might affect each person differently. One needs to devise an adaptive plasma approach to achieve the desirable outcome in using CAP. To that end, implementing real-time monitoring of the tissue/body response is essential and should be considered as a priority.

The new platform described by Keidar *et al* [11] is intended for monitoring the cellular response to CAP in a continuous read method, with the RealTime-Glo assay serving as part of the overall feedback system. The data collected will be analyzed by a feedback algorithm relying on viability measurements of healthy and cancerous cells to predict the optimal amount of (1) delivery power and (2) exposure time. These two parameters will regulate the CAP intensity and the amount of RONS delivered to cells during treatment. The feedback control algorithm will be developed according to nonlinear model-predictive controls (MPCs). MPC or receding-horizon control are used to implement numerical solutions of open-loop optimization in the form of feedback. MPC is particularly desirable for the proposed adaptive plasma platform, as the treatment conditions will be determined in an optimal fashion to maximize the decay rate of the tumor. The algorithm has been shown to leverage a model of plasma–tumor interactions. The feedback system will employ the adaptive properties of CAP to yield optimal and appropriate treatments tailored to real-time diagnostics while laying a foundation for personalized adaptive cancer therapy.

10.3 MPC-based adaptive plasma platform

Various factors, such as plasma jet discharge voltages, gas composition, humidity, and type of cancer cells influence CAP effect on cancer cells. As such, it is challenging to plan plasma cancer treatments for a particular cancer in the clinical setting. To address this issue, we present an optimal feedback control scheme to adjust treatment conditions responsive to the actual cancer cell response.

Recently Gidon *et al* presented control schemes for a device that generates an atmospheric pressure plasma jet to control the target's temperature [15, 16]. To that end, a model-predictive control strategy was proposed for real-time feedback control of a radio-frequency plasma discharge in argon. The closed-loop performance of the MPC strategy was compared to that of a basic proportional–integral control system. It was indicated that the MPC strategy provides a versatile framework for dose delivery in the presence of disturbances.

Lyu *et al* [17] developed a mathematical model for the dynamic response of cancer cells to CAP jets, based on experimental data that provide temporal measurements of cancer cell viability after CAP treatments. A differential equation is developed to model the influence of the CAP on the viability of U87 and

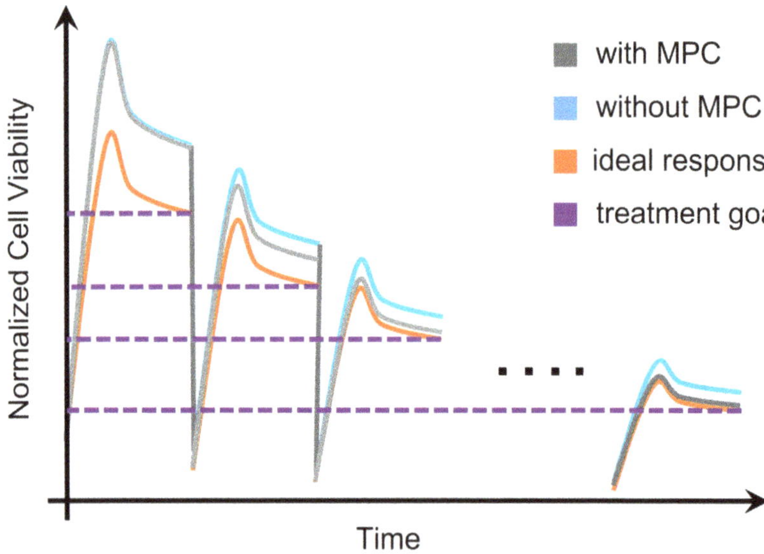

Figure 10.5. The idea of treating cancer cells using MPC versus no-MPC [17].

MDA-MB-231 *in vitro* under varying treatment durations and plasma discharge voltages. A control scheme is presented to determine CAP treatment conditions optimally to reduce cancer cell viability to less than a prescribed goal while minimizing a weighted sum of the treatment duration and the discharge voltage. This is further extended to a model-predictive control framework such that a preplanned CAP treatment schedule is revised according to the actual response of cancer cells [16]. The efficacy of the proposed approach is illustrated by numerical simulations based on experimental data.

The corresponding simulation results for U87 can be found in reference [17], and the general idea is illustrated in figure 10.5. The purple dash lines show the desired relative cell viability at the end of each period. The orange solid curves represent the ideal case when the mathematical model for the actual response is utilized. In this ideal case, the cell viability is reduced to the desired level at the end of each period. However, when the mathematical model available for optimal control is not identical to the model for the actual response, as illustrated by blue curves, the cell viability becomes greater than the desired level, greater than the desired value. Finally, when the proposed MPC is introduced, the controlled cell viability reduces to the desired level, even with the discrepancy in the mathematical modeling. As such, this simulation result suggests that adjusting CAP treatment conditions adaptive to the cancer response can mitigate the adverse effects of modeling errors.

10.4 Electrochemical response of cells to plasma: real-time impedance measurements

In addition, various ways to supplement the cell viability for real-time diagnostics (such RealTime™ assay mentioned above) are considered. This includes the

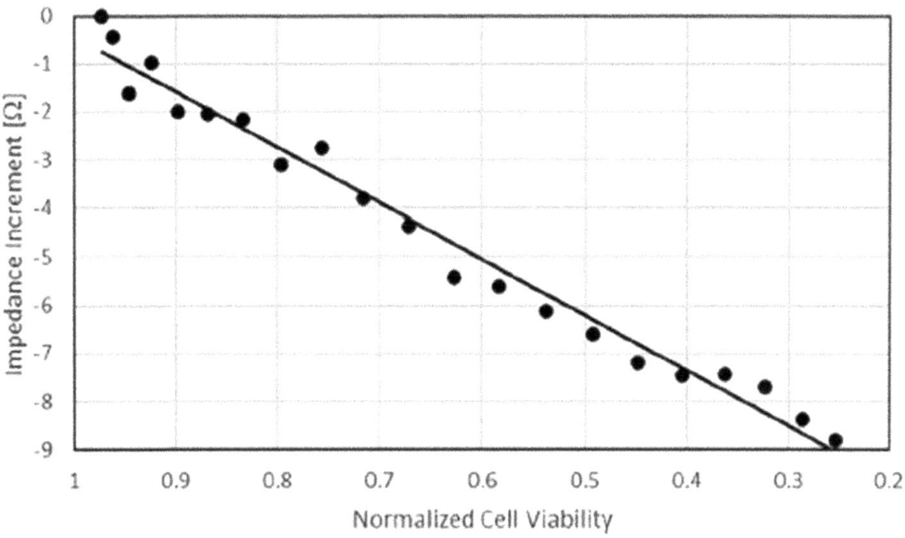

Figure 10.6. Correlation between impedance and cell viability. Reprinted from [18], with the permission of AIP Publishing.

electrochemical impedance spectroscopy (EIS) that can provide a real-time feedback signal for MPC method. The EIS frequency range may vary from kHz to GHz [18]. The equivalent circuit combines resistors, capacitors, inductors, and nonideal circuit elements [19], such as constant phase elements. Recent results indicate that the electromagnetic emission from the CAP may enable the cells into an activation state that is more sensitive to reactive species and medicine [20, 21]. One of the physical explanations of such a phenomenon is the membrane vibration due to the charges on the membrane oscillating in the incoming electromagnetic emission. The membrane vibration thus produces an acoustic wave propagating through the intracellular substance. Such a theory may be verified using the EIS α dispersion for the membrane vibration and γ dispersion for the intracellular acoustic wave. The correlation between EIS signal and cell viability (MTT assay) is shown in figure 10.6 [18]. Such a correlation might be used in the feedback guidelines of adaptive plasma.

10.5 Adaptive plasma treatment based on machine learning technologies

The adaptation of plasma can also focus on any of the chemical species in the plasma rather than the cell treatment effects. However, the complexity of plasma chemistry in air is substantial, which contains thousands of chemical reactions and hundreds of chemical species, including ions, photons with different wavelengths, and excited species at different states [23–25]. Therefore, the most straightforward automatic control and optimization of the plasma treatment system for specific chemical species or parameters such as gas bulk temperature is to use a machine

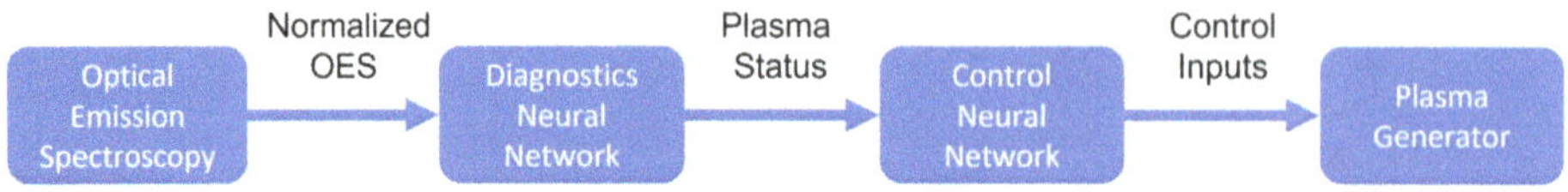

Figure 10.7. The flow chart of a two-ANN adaptive plasma system.

learning system [25–27]. For controlling the temperature in real time, the temperature on target or in the plasma plume can be measured by any temperature meter and provides a feedback signal to the plasma generator. For example, a microcontroller carrying artificial neural networks can control the plasma jet power in real time and lead to the temperature on target following a desired variation over time. After the training, the adaptive system clearly understands how to manipulate the target temperature [26].

However, it won't be easy to provide the complete plasma species composition in real time when one needs to adapt the plasma for a specific species concentration. Therefore, machine learning can contain two artificial neural networks (ANNs) in series. As shown in figure 10.7, the first ANN will translate the real-time optical emission spectrum (OES) to the species composition map, and the second one will thus provide control, including the gas supply, discharge voltage, etc, to the plasma generator until the targeting species concentration in the composition map fulfills the requirement. The training of these ANNs is not trivial due to the complex relationship between the plasma generator's OES and control parameters. Therefore, an evolutionary training algorithm should be used instead of the backpropagation method of ANNs [25].

The adaptive plasma can finally be a general device which can be switched among different applications, such as maximizing OH for cancer therapy and O_3 for sterilization. Each application requires a specific training result, a.k.a. the weight matrix, which can be saved in the microcontroller chip in the plasma generator [25]. With such an adaptation focusing on specific chemical species under real-time control using machine learning, the plasma can also be immune to environmental disturbance and always keep the required species at the wanted concentration [25].

10.6 Intelligent plasma based on self-organization patterns

A more advanced method uses a novel concept called 'intelligent plasma'. Intelligent plasma is a newly discovered type of intelligent material, which reflects two key features that are different from other intelligent materials on the market: programmable and machine learning of high-level intelligence [28]. The idea is to consider the chemical pathway network (CPN) in a plasma system where each of the species' concentrations are the neurons, reaction rate coefficients are the activation functions, and the parameters controlling the coefficients, such as reactant temperatures, are the weights. Therefore, some species' concentrations can be used as the input neurons while others can be used as the outputs. Combining this concept with the pre-mentioned self-organization pattern can achieve a more advanced type of adaptive plasma that can avoid the normal cells but focus the plasma plume on cancer cells.

First, several recent papers reported that the streamer space charge near the target depends on the target status [29–31] and thus the plasma jet can deform and incline [32]. The physical mechanism behind this effect is the following. The streamer's electric field is determined by the ion density in the streamer head and the target surface when approaching the target. Therefore, when the boundary condition, such as the electric potential value, is fixed and is lower than the streamer potential, a higher target permittivity (dielectric constant) can lead to a stronger electric field before the streamer head. This helps the streamer to propagate. When two or more targets with different permittivity values are set side-by-side for the plasma jet to treat, the permittivity distribution will thus determine where the streamer propagates, resulting in a brightness distribution known as the self-organization pattern. The cell-medium dielectric constant and capacitance of the cancer cells are both higher than the values of normal cells, as shown in figure 10.8. This provides a possibility that the plasma jet can avoid normal cells and focus on cancer cells, as shown in figure 10.9.

Therefore, the intelligent plasma concept applied to controlling the self-organization of the plasma jet by itself can be summarized in figure 10.10. The intelligent plasma, a.k.a. plasma computer, can always generate proper self-organization patterns according to the target status represented by the gas supply, under a constant set of CPN weights (generator setup parameters such as the target distance, discharge voltage, etc). Therefore, to make the system practical, one needs the following two parts. (1) A protocol to translate the targeting cancer-normal cell distribution to the gas supply. This can be done by measuring the dielectric constant distribution in the dish to find the cancer-normal cell distributions. For example, by applying AC voltages between each neighboring electrode which is equivalent to a capacitor, an array of pad electrodes beneath the dish, the impedance between them will indicate the dielectric constant in that area. Then, the gas supply valves will be controlled by a

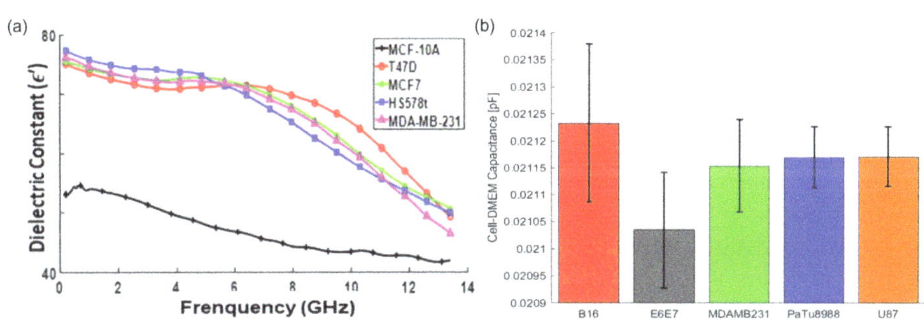

Figure 10.8. Cell-medium dielectric constants and cell-medium capacitance [18], where MCF-10A is breast epithelial cell, MDA-MB-231, Hs578T, MCF-7 and T47-D are breast cancer cells, B16 is skin cancer cell, E6E7 is normal astrocyte, PaTu8988 is pancreatic adenocarcinoma cell, and U87 is the brain cancer cell. (a) Reproduced from [22]. CC BY 4.0. (b) Reprinted with permission from [32], copyright (2019) American Chemical Society.

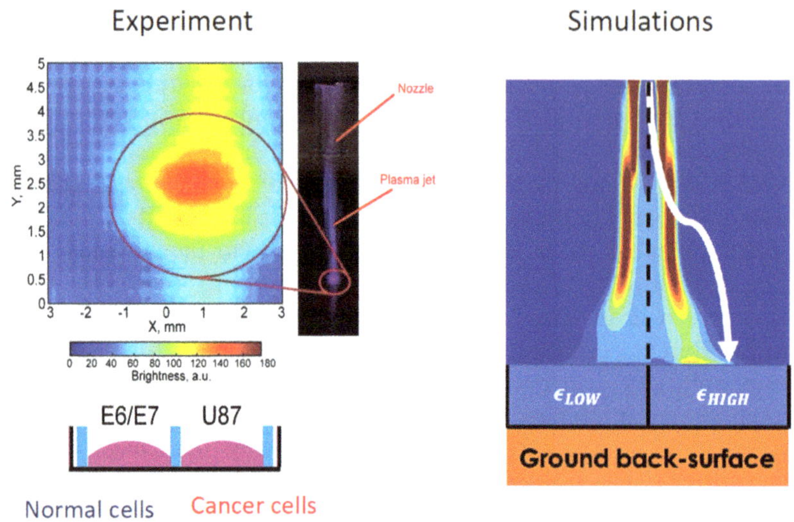

Figure 10.9. Plasma jet self-organization at the interface with cancer and normal cells. The plasma jet is shifted toward cancer cells. Simulations confirm that self-organization results from the dielectric properties difference between cancer and normal cells.

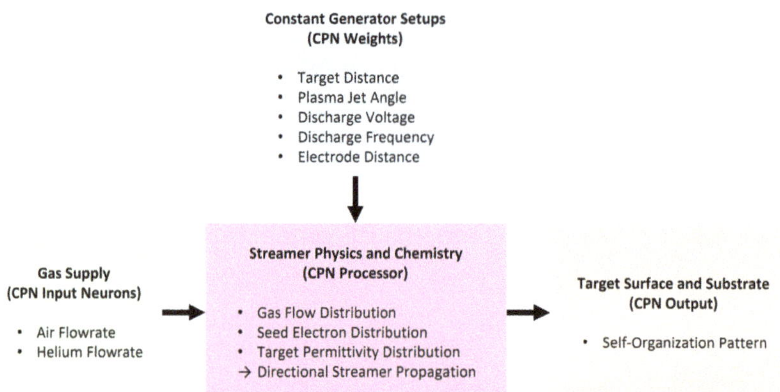

Figure 10.10. The concept of self-organization pattern generated by intelligent plasma, where one can find the von Neumann architecture of such a 'plasma computer'.

microcontroller according to the pad electrode signals. (2) The training of CPN to find the proper constant values of CPN weights can always make the intelligent plasma jet produce the self-organization pattern with acceptable accuracy. In training, one considers the CPN as a regular ANN, and the training can be done automatically by a microcontroller connected to the plasma generator without real cells. First, the training dataset is generated randomly in the microcontroller to mimic the real cell distributions. The microcontroller will then feed the gas supply according to the virtual cell distribution to the plasma jet generator. After a digital camera captures a self-organization pattern, the image will be sent back to the microcontroller to compute the error between the image and the virtual cell distribution. The error will then be used to update the weights of CPN based on the evolutionary algorithm like the one introduced in the previous section.

References

[1] Ross A W 2004 Principles of the self-organizing system *ECO Emerg. Complex. Organ.* **6** 102–26

[2] Nicolis G and Prigogine I 1977 *Self-Organization in Nonequilibrium Systems* (New York: Wiley)

[3] Benilov M S 1992 Theory of structures in near-electrode plasma regions *Phys. Rev.* A **45** 5901

[4] Kogelschatz U 2002 Filamentary, patterned, and diffuse barrier discharges *IEEE Trans. Plasma Sci.* **30** 1400–8

[5] Shi J J, Liu D W and Kong M G 2006 Plasma stability control using dielectric barriers in radio-frequency atmospheric pressure glow discharges *Appl. Phys. Lett.* **89** 081502

[6] Akishev Y, Goossens O, Callebaut T, Leys C, Napartovich A and Trushkin N 2001 The influence of electrode geometry and gas flow on corona-to-glow and glow-to-spark threshold currents in air *J. Phys. D: Appl. Phys.* **34** 2875–82

[7] Laroussi M, Alexeff I, Richardson J P and Dyer F F 2002 The resistive barrier discharge *IEEE Trans. Plasma Sci.* **30** 158–9

[8] André P, Barinov Y, Faure G, Kaplan V, Lefort A, Shkol'Nik S and Vacher D 2001 Experimental study of discharge with liquid non-metallic (tap-water) electrodes in air atmospheric pressuerr *J. Phys. D: Appl. Phys.* **34** 3456

[9] Trelles J P 2016 Pattern formation and self-organization in plasmas interacting with surfaces *J. Phys. D: Appl. Phys.* **49** 393002

[10] Chen Z, Zhang S, Levchenko I, Beilis I I and Keidar M 2017 In vitro demonstration of cancer inhibiting properties from stratified self-organized micro-discharge plasma–liquid interface *Sci. Rep.* **7** 12163

[11] Keidar M, Yan D, Beilis I I, Trink B and Sherman J H 2018 Plasmas for treating cancer: opportunities for adaptive and self-adaptive approaches *Trends Biotechnol.* **36** 586–93

[12] Gjika E, Sonali P G, Anna T, Megan K, Gauri T, Jerome C, Mary A S and Keidar M 2018 Adaptation of operational parameters of cold atmospheric plasma for *in vitro* treatment of cancer cells *ACS Appl. Mater. Interfaces* **10** 9269–79

[13] Keidar M 2016 Therapeutic approaches based on plasmas and nanoparticles *J. Nanomed. Res.* **3** 00052

[14] Keidar M 2018 A prospectus on innovations in the plasma treatment of cancer *Phys. Plasmas* **25** 083504

[15] Gidon D, Graves D B and Mesbah A 2017 *Plasma Sources Sci. Technol.* **26** 085005

[16] Gidon D, Curtis B, Paulson J A, Graves D B and Mesbah A 2017 Model-based feedback control of a kHz exited atmospheric pressure plasma jet *IEEE Trans. Radiat. Plasma Med. Sci.* **2** 129–37

[17] Lyu Y, Lin L, Lee T and Keidar M 2019 Mathematical modeling and control for cancer treatment with cold atmospheric plasma jet *J. Phys. D: Appl. Phys.* **52** 185202

[18] Lin L, Hou Z, Yao X, Liu Y, Sirigiri J R, Lee T and Keidar M 2020 Introducing adaptive cold atmospheric plasma: the perspective of adaptive cold plasma cancer treatments based on real-time electrochemical impedance spectroscopy *Phys. Plasmas* **27** 063501
[19] Lvovich V F 2012 *Impedance Spectroscopy Applications to Electrochemical and Dielectric Phenomena* (Hoboken, NJ: Wiley)
[20] Yan D, Xu W, Yao X, Lin L, Sherman J H and Keidar M 2018 The cell activation phenomena in the cold atmospheric plasma cancer treatment *Sci. Rep.* **8** 15418
[21] Yan D, Lin L, Xu W, Nourmohammadi N, Sherman J H and Keidar M 2019 Cold plasma-based control of the activation of pancreatic adenocarcinoma cells *J. Phys. D: Appl. Phys.* **52** 445202
[22] Hussein M, Awwad F, Jithin D, El Hasasna H, Athamneh K and Iratni R 2019 Breast cancer cells exhibits specific dielectric signature *in vitro* using the open-ended coaxial probe technique from 200 MHz to 13.6 GHz *Sci. Rep.* **9** 4681
[23] Norberg S A, Johnsen E and Kushner M 2015 Formation of reactive oxygen and nitrogen species by repetitive negatively pulsed helium atmospheric pressure plasma jets propagating into humid air *Plasma Sources Sci. Technol* **24** 035026
[24] Lin L, Lyu Y, Trink B, Canady J and Keidar M 2019 Cold atmospheric helium plasma jet in humid air environment *J. Appl. Phys.* **125** 153301
[25] Lin L, Yan D, Lee T and Keidar M 2022 Self-adaptive plasma chemistry and intelligent plasma medicine *Adv. Intell. Syst.* **4** 2100112
[26] Witman M, Gidon D, Graves D B, Smit B and Mesbah A 2019 Sim-to-real transfer reinforcement learning for control of thermal effects of an atmospheric pressure plasma jet *Plasma Sources Sci. Technol.* **28** 095019
[27] Dogan G, Graves D B and Mesbah A 2019 Predictive control of 2D spatial thermal dose delivery in atmospheric pressure plasma jets *Plasma Sources Sci. Technol.* **28** 085001
[28] Lin L and Keidar M 2022 Artificial intelligence without digital computers: programming matter at a molecular scale *Adv. Intell. Syst.* **4** 2200157
[29] Norberg S A, Johnsen E and Kushner M J 2015 Helium atmospheric pressure plasma jets touching dielectric and metal surfaces *J. Appl. Phys.* **118** 013301
[30] Schweigert I V, Vagapov S, Lin L and Keidar M 2019 Increase of electric field on streamer/surfaces contact area with using external electrode *J. Phys.: Conf. Ser.* **1394** 1012005
[31] Schweigert I V, Vagapov S, Lin L and Keidar M 2019 Enhancement of atmospheric plasma jet–target interaction with an external ring electrode *J. Phys. D: Appl. Phys.* **52** 295201
[32] Lin L, Yan D, Gjika E, Sherman J H and Keidar M 2019 Atmospheric plasma meets cell: plasma tailoring by living cells *ACS Appl. Mater. Interfaces* **11** 30621–30

www.ingramcontent.com/pod-product-compliance
Ingram Content Group UK Ltd.
Pitfield, Milton Keynes, MK11 3LW, UK
UKHW051853100926
8556IPUK00005B/598

9 780750 355353